Lecture Notes in Physics

Lecture Notes in Physics

123

Dieter H. Mayer

The Ruelle-Araki Transfer Operator in Classical Statistical Mechanics

Springer-Verlag
Berlin Heidelberg GmbH 1980

Author

Dieter H. Mayer
Institut für Theoretische Physik E
RWTH Aachen
Sommerfeldstraße
D-5100 Aachen

ISBN 978-3-540-09990-1   ISBN 978-3-540-39280-4 (eBook)
DOI 10.1007978-3-540-39280-4

Library of Congress Cataloging in Publication Data. Mayer, Dieter H 1942- The Ruelle-Araki
transfer operator in classical statistical mechanics. (Lecture notes in physics; v. 123)
Bibliography: p. Includes index. 1. Statistical mechanics. 2. Linear operators. I. Title.
II. Series. QC174.86.C6M39 530.1'32 80-14968

© by Springer-Verlag Berlin Heidelberg 1980
Originally published by Springer-Verlag Berlin Heidelberg New York in 1980

2153/3140-543210

à CHRISTIANE

PREFACE

Modern statistical mechanics started at the beginning of this century with the remarkable work of the American mathematical physicist J. Willard Gibbs. His monograph on "Elementary Principles in Statistical Mechanics" which appeared in 1902 marks a milestone in the conceptual clarification of the principles of statistical mechanics. For more than half a century this book served as the standard reference for all those who saw in this discipline more than only a collection of prescriptions for calculating macroscopic quantities.

A new dimension in the discussion of the mathematical structures underlying statistical mechanics was certainly opened in 1968 when David Ruelle's book on "Statistical Mechanics: Rigorous Results" was published. Without being a prophet one can say that this monograph will play the same role for the rest of the century as did Gibb's book for the first half.

Since the appearance of Ruelle's book rigorous statistical mechanics has developed rapidly. It has become more and more clear that the structures underlying this theory are of a much more general nature and can be found also in completely different domains like the theory of general dynamical systems.

This relation between an old physical theory such as statistical mechanics on the one side and abstract mathematical systems on the other seems to be a common feature in our times: one remembers also the recently found connection between the modern gauge field theories of elementary particle physics and problems of algebraic geometry. All this can lead to a new era of intensive discussions among physicists and mathematicians on problems of interest to both.

In the present work we discuss in a very special example the intimate relation between a physical system and its underlying mathematical structure. From the work of Sinai and Ruelle it has become obvious that one-dimensional classical lattice systems are of great interest for both physicists and mathematicians. They form the basic structure of a wide class of dynamical systems and have contributed greatly to the understanding of the behaviour of such systems.

Of particular interest are such one-dimensional lattice systems with long range interactions. Unfortunately however, the mathematically rigorous description of these systems is not yet in a completely satisfactory status when compared with that for systems with finite range interactions. The first steps toward such a description were made by Ruelle and Araki who constructed generalized transfer matrices for the case of long range interactions. The power of this method can, however, not yet be compared with the classical transfer matrix method for finite range interactions.

We will discuss in this work how the method of Ruelle and Araki can be improved, at least for certain classes of interactions, to become completely equivalent to the former one. This leads to interesting mathematical problems in the spectral theory of certain linear operators.

I started this work during a visit in the years 1975/1976 at the Institut des Hautes Etudes Scientifiques in Bures sur Yvette in France financed by a fellowship of the Deutsche Forschungsgemeinschaft.

My special thanks are due to Prof. D. Ruelle who introduced me to the problems described in this work. I had the pleasure of following his lectures at the mathematics department at the University of Orsay on "Thermodynamic Formalism" which later appeared in book form, in which a much deeper discussion of the structures inherent in statistical mechanics is presented.

Thanks are also due to Prof. K. Viswanathan from Simon Fraser University for fruitful collaboration over the past years and to Prof. G. Roepstorff from the RWTH Aachen for many discussions on the ideas described in this work.

Aachen, February 1980   Dieter H. Mayer

CONTENTS

INTRODUCTION AND RESUMÉ

The aim of statistical mechanics is to explain the macroscopic measurable properties of a system composed of a large number of constituents starting from the dynamical laws and principles valid for these microscopic constituents in general. A question which arises quite naturally in this connexion is why all kind of matter we know appears to exist in exactly three different phases namely as a gas, a liquid or a solid. Unfortunately, we are still a long way from being able to give a convincing and well founded answer to this and similar fundamental questions.

The difficulties arising in understanding such problems seem to be strongly correlated with the dimension of the system we are looking at: whereas systems in one or two dimensions have seen more or less important progress in an exact treatment over the last years this seems not to be the case in three dimensions . Unfortunately however this is just the dimension where nature is used to live. There does not exist any model in dimension higher than two which would show interesting physical behavior and which could be solved exactly.

Certainly, it is not always absolutely necessary to know the exact solution of a system to understand certain of its properties. But without doubt this would be the ideal way to discuss and understand a system completely. In one and two dimensions there exist indeed models which can be solved exactly and where the relevant physical quantities can be written down in the form of analytic expressions. The most famous example for such a system is certainly Onsagers solution of the two dimensional Ising system with nearest neighbour interaction and vanishing exterior magnetic field.

Unfortunately however it is not known how far the mechanisms responsible for the behavior of such low dimensional systems are still valid in higher dimensions. Nature is as we said already mostly three-dimensional and any one-or two-dimensional model for it can give only an approximate and presumably very crude description of what is really going on. Therefore it is not at all clear if we can really learn something from such low dimensional systems in our desire to understand such physical phenomena as the above mentioned phase transitions in real matter.

Nevertheless we think that the study of such one-and two-dimensional systems can be of some interest . These systems provide us with a possibility to test new ideas and methods which one hopes to apply finally also to real three dimensional models. If one knows the exact solution of a system then it is in general very easy to see the limitations and problems arising with new methods which one cannot in general guess from a priori. From this point of view also lower-dimensional systems merit therefore a detailed study.

The best understood systems are certainly the lowest-dimensional ones- the systems in one dimension. This is true for discrete and continuous systems both classical and quantum mechanical, both in equilibrium and non-equilibrium. The simple reason for this seems to be connected with the poor geometrical freedom of such systems: extended objects can never change a given order in their arrangement on a one-dimensional line. This order stays the same for all times,respectively it limits the allowed phase space in such a way to enable exact solutions. The mathematical difficulties increase in fact tremendously as soon as such objects can pass each other in moving around in further dimensions.

Unfortunately these one-dimensional models have one drawback which make them not very appealing from the physical point of view. It turns namely out that these systems are much poorer in their physical behaviors then the higher dimensional systems. In fact,as long as one is not willing to accept very long range interactions which anyhow appear not to be very natural it was shown already very early[1],[2]that such one dimensional models are described by analytic thermodynamic functions. But this is only the mathematical way for saying that such systems do not have phase transitions which are always connected with some non smooth behavior of some physical observable.

It is therefore not immediately obvious why one is nevertheless interested in these one dimensional systems. One point which we mentioned already also in the case of two dimensional models and which certainly is even more convincing for one-dimensional models is that such models constitute an interesting testing ground for new approaches and ideas. Because of their simplicity one hopes to understand the validity and limitations of such new ideas faster and easier than in higher dimensions. This strategy is well known and has been applied in fact very often [3].

An interesting example which should be mentioned in this connexion is the so called renormalization group method originally invented by Kadanoff [4] and Wilson [5]. This method is known to give very good results in the phenomenological description of the so called critical phenomena both in statistical mechanics and in quantum field theory. But so far it is not really understood why this works all so well.

It was only very recently that one succeeded in giving a mathematical well founded formulation and description of this method. And this was done in fact for simple classical spin systems on a one-dimensional lattice [6],[7] which have phase transitions because of the long range nature of their interactions.

The examples where low-dimensional systems served as milestones for understanding new ideas and methods and also for new developments are certainly not restricted to the area of statistical mechanics. We mention only the important developments made just over the last ten years or so in understanding the problems arizing in quantum field theories through a detailed study of two-and three-dimensional models [8] , [9]. Because one dimension is just ordinary time a two-dimensional model there corresponds in fact to an one-dimensional model in equilibrium statistical mechanics.

A comprehensive review of the whole area of one dimensional physics can be found in [1] which is still worth reading even if the latest developments in this field are not discussed there.

The last ten years saw a growing interest in one-dimensional systems of statistical mechanics from a completely new point of view. In a series of papers it was shown by Sinai [10]-[12] that there exists an interesting relation between these systems and the abstract mathematical theory of dynamical systems. He found that in many of such abstract systems there arizes a structure completely similar to the one of our classical spin systems on a one-dimensional lattice with certain long range interactions. This relation allowed Sinai and later also Bowen and Ruelle [13]-[22] to solve many measure theoretic problems for a wide class of such dynamical systems by translating known results of the physical models into the language of mathematics.

Starting from these papers there originated a completely new direction of research in mathematical physics known today under the name "thermodynamic formalism" [22]. There one tries to understand the mathematical basic structure underlying the principles of statistical mechanics and thermodynamics and to generalize this to more abstract mathematical systems for solving problems arising in the discussion of such dynamical systems.

This development on the other hand is of course of great interest not only from a mathematical point of view: in fact some and indeed the most fundamental open problems in statistical mechanics are closely related to the theory of dynamical systems : recall only the problem of the foundations of both equilibrium and non-equilibrium statistical mechanics [23], [24].

It turns out that this abstract general theory which serves also as a global theory of differential equations [25] can make more and more

very important contributions to our understanding of a variety of physical phenomena which have been open problems for many years. Let us
mention in this connection only the recent developments in the theory
of the turbulent motion of viscous fluids. The works of Lorenz [26]
and Ruelle and Takens [27], [28] have brought new ideas   in problems
which are unsolved for almost a century now. We mention also the applications of this theory in the fields of biology or chemical reaction
theory [29]. Even if this looks at the first moment a little bit
surprizing such a complicated system like a fluid seems to have at its
basis a mathematical structure similar to the one of a one-dimensional
discrete spin system on a lattice [30] !

Having in mind such far reaching connections it should be clear
that one-dimensional systems are interesting objects and worth a detailed study. This is especially necessary for those with long range
interactions which are still far from being understood in a sufficient
way. On the other hand such systems with finite range interactions
are well understood since quite a long time. Both the discrete and
the continuous systems can be described in this case by thermodynamic
functions which are analytic in all relevant parameters like temperature, pressure and so on [31] - [33].

For systems with long range interaction the situation is a little bit different. Let us mention a few of the problems which one
would like to understand better.

There is the problem of finding necessary and sufficient conditions on the interaction for the existence of a phase transition. The
old conjecture of Dyson [35] is neither proved nor disproved. It says
that in an Ising system [36] with an interaction depending on the distance as $J(i)$ there exists a phase transition in the form of a spontaneous magnetization if, and only if this function $J(i)$ fulfills the
following two conditions:

1) $\displaystyle\sum_{i=1}^{\infty} |J(i)| = \infty$ and 2) $J(i) \geq c \; 1/i^2$ for $i \to \infty$,

where c is some constant [37]. Especially the limiting case $J(i) \sim i^{-2}$

is still very controversial also regarding the possible order of a

phase transition [38] - [40].

Dyson's conjecture is more or less concerned with phase transi-

tions of the first order that means those accompanied by a spontaneous

symmetry breakdown. Higher order phase transitions [41] are touched

only by the Anderson model $J(i) = i^{-2}$. Such phase transitions are

defined by the analyticity properties of the thermodynamic functions.

Only in the case these functions are real analytic in the physical re-

gion of all the parameters one can say the system has no phase transi-

tion at all.

For some time the belief was [42] that the order n of such a pha-

se transition is determined by the smallest number k for which the k-th

moment $\displaystyle\sum_{i=1}^{\infty} |J(i)| i^k$ of the function $J(i)$ diverges.

For n=0 and n=1 this belief was confirmed by Dyson [43] - [44] and

Fisher [45] though in a little bit modified form .

If one applies the above conjecture to interactions decreasing

exponentially fast with the distance i one expects at most an infinite

order phase transition, that means all the thermodynamic functions

should be infinitely often differentiable in the relevant parameters.

And this was in fact shown to be the case by Araki [46] for a one-dim-

ensional quantum system. He showed even more:namely the free energy

of such a system is a real analytic function in the temperature and

the other parameters. Ruelle [47] derived later a completely analogous

result for classical spin systems on a one-dimensional lattice.

Another class of models with an interaction J behaving like

$J(i) \sim \exp-\gamma \; i^{\beta}$, $0 < \beta < 1$, $\gamma > 0$ was discussed by Gallavotti and Lin

[42] . These interactions are quite interesting because they decrease
at infinity slower than exponentially but still faster than any polynomial. The above authors showed that the thermodynamic functions in
this case are infinitely differentiable, but it is not known if they
are also real analytic like this is the case for exponentially decreasing interactions.

It was therefore a surprize when Dobrushin showed [48] that even
for polynomially decreasing interactions the free energy respectively
the correlation functions are real analytic as long as the function
$J(i)$ behaves like i^{-m} with $m \geq 3$ at infinity. The techniques used
by Dobrushin originate in the theory of stochastic processes and are
therefore completely different from those used by the authors mentioned
before. They used methods based more or less on a generalized transfer matrix method invented in its classical form by Kramers and Wannier
[49], Montroll [50] and for one dimensional systems by Ising [36] . This
method consists in shifting the problem of calculating the partition
function for a system to an algebraic problem, namely the determination
of the spectrum of a matrix or in general a linear operator which can
be attributed to such a system.

It arises therefore the question if it is possible to derive the
results of Dobrushin also by this transfer matrix method. This seems
to be quite desirable in view of the fact that Dobrushin's proof is
not very easy to understand [47] .

There arise however immediately some difficulties. Originally,
the transfer matrix method was invented for systems with finite range
interactions. Later Araki [46] and Ruelle [51] respectively Gallavotti
and Miracle-Sole [52] succeeded in defining what they called a "generalized transfer matrix" also for long range interactions. This "matrix"
is defined as a certain linear operator in the space of observables
of such systems. The justification for the name "generalized trans-

fer matrix" stems from the fact that the spectrum of this operator, to
be more precise the highest eigenvalue is related to the free energy
of the system in much the same way as this is the case for finite range
interactions for the Kramers Wannier matrix. However the spectrum of
these generalized transfer matrices is in fact much more complicated
and therefore also much less understood so that this method allowed
so far to derive only much weaker results for such systems with long
range interactions when compared with the classical transfer matrix.
In the latter case one knows in general the complete spectrum of the
transfer matrix, which anyhow for discrete systems is a finite-dimen-
sional matrix, whereas for the Ruelle-Araki transfer matrix only some
properties of the highest eigenvalue are more or less known. There
is no example where this highest eigenvalue is in fact explicitly known
besides the proof that it exists. This kind of information is certainly not enough to draw further conclusions about for instance the analytic behavior of the free energy.

To prove something in this direction a much deeper knowledge of
the spectral properties of these operators in the neighbourhood of the
largest eigenvalue would be necessary to get statements about differentiability and so on [53].

For polynomially decreasing interactions one could show so far
by this method only that the thermodynamic potentials are once differentiable in the temperature and all other parameters.

In the case of exponentially decreasing interactions the situation is much better. For systems with an interaction where $J(i)$ beha-
ves like $a(i)$ exp-γi^{β} with $\gamma > 0$ and $\beta \geq 1$ Ruelle improved the above
result and showed that the free energy in such a case is in fact real
analytic. He got this result by proving that the highest eigenvalue
of the generalized transfer matrix is simple and completely separated
from the rest of the spectrum of this operator. From this real analyticity of the free energy follows then immediately [54].

But even in this case Ruelle did not get by his method any result about the rest of the spectrum. As we know however such results
would be necessary for a complete description of a system for instance
by the correlation functions [55] which in fact is provided by the classical transfer method for finite range interactions.

Also a simple method is missing how to calculate the highest eigenvalue for a given interaction Φ .

These problems were the point of departure for our investigations described in this work. We posed ourselves the question how far
can the method initiated by Ruelle and Araki be improved to give finally a description for systems with such long range interactions which
is as efficient as the old transfer matrix method of Ising, Kramers
and Wannier for finite range interactions. This includes in particular
also the aim of getting a reasonable knowledge of the spectrum of these
generalized transfer matrices and a simple method for calculating at
least in principle the relevant quantities of interest.

We can say that we achieved this goal more or less for exponentially decreasing interactions. These interactions can be described by
the Ruelle-Araki transfer matrix in a way which is completely equivalent to the one provided by the Kramers-Wannier matrix for finite range interactions.

Unfortunately we did not succeed to extend the method of Ruelle
and Araki in such a way that also polynomially decreasing interactions
can be treated in the same way. This remains a challenging problem
and is in fact much more interesting than what we can present here.

The same is true also for the systems discussed by Gallavotti and
Lin where the function $J(i)$ behaves like $\exp-\gamma i^{\beta}$ with $0 < \beta < 1$.
Only the case $\beta \geq 1$ can be treated by our method in a satisfactory way
so far. Concerning the last cases we present only some ideas how in

our opinion a solution of the aforementioned problems could possibly be found. We are convinced that the mathematical problems we are facing in the general case of a polynomially decreasing interaction can be solved some day.

The method which we want to discuss in this paper is a straight-forward extension of the method of Ruelle and Araki and relies on a very simple mathematical idea: given an operator in some large space with a complicated spectrum one could first of all try to study this operator, if possible, in some smaller invariant subspaces of the original space. This way one can hope to get by a clever choice of this invariant subspace an operator which has on the one hand a simpler spectrum but, on the other hand, is still capable to describe also all the physical properties of the system under consideration as it did the operator in the large space. Unfortunately there does not exist a general theory how one should do this. So the success of chosing in a concrete case this and this subspace is more or less a matter of chance. It depends very much on the ability and the insight of the single person.

In this sense one could formulate our results as follows: in the case of exponentially decreasing interactions we found the right subspaces for the Ruelle-Araki operator to become the real transfer matrix for the systems whereas in the other cases, especially for polynomially decreasing interactions, we did not yet find the right spaces.

Unfortunately we do not know if the reason for this is of a more fundamental nature or we have been only not clever enough so far. But the results of Dobrushin in our opinion are some indirect indications that our method could finally work also for the last mentioned systems.

To make the reader acquainted with the above quite general idea we will apply it first in the case of finite range interactions. We explain how our formulation of the Ruelle-Araki transfer matrix reduces this operator to the classical Kramers-Wannier transfer matrix if the above described method of restricting the domain of definition of the former is applied in an appropriate manner. For reasons of completeness we recall this Kramers-Wannier matrix for simple systems with finite range interactions in the first chapter where we introduce also our notations and definitions.

In the second chapter we consider general properties of the Ruelle-Araki operator and discuss the above mentioned relation of this operator to the Kramers-Wannier matrix. As an application of these ideas we derive a transfer matrix for the continuous hard rod system with finite interaction in the grand canonical ensemble. Such a matrix was considered so far only in the so called pressure ensemble going back to van Hove.

Chapters III. and IV. constitute the real heart of this work. There we treat exponentially decreasing interactions. We show how our method allows a detailed discussion of the spectral properties of the Ruelle-Araki operator for such interactions. This operator becomes the real transfer matrix of such systems by restricting its domain of definition to certain Banach spaces of holomorphic functions. In these spaces the operator becomes in fact a nuclear operator in the sense of Grothendieck. We derive also simple formulas which allow us to calculate the eigenvalues of this operator at least in principle. This we then apply to a simple model introduced by Kac which allows a rigorous derivation of the van der Waals equation of state respectively of the Weiss theory of ferromagnetism.

After the discrete spin systems we discuss continuous spin systems. Their treatment is not any more so simple as the one of the discrete systems and leads us already to certain limitations of our method in the present form.

The problems arising in the discussion of the continuous spin sys-

tems are similar to the ones for discrete spin systems with an interaction where the function J(i) behaves at infinity like exp-γ i$^\beta$ with
$\gamma > 0$ and $\beta \geqslant 1$. But contrary to the former case we can resolve the
problems in the latter one. The mathematical theory which allows us
this solution is again Grothendieck's theory of nuclear operators in
general Banach spaces.

In the last chapter we will use the results obtained in the previous chapters to improve the analyticity properties of the so called
zeta-functions for such one-dimensional classical systems with exponentially decreasing interactions. These functions have been introduced into statistical mechanics by Ruelle in analogy to such functions
in the theory of dynamical systems. They provide an elegant mathematical description of such systems. In this chapter we touch therefore
the interesting relation between statistical mechanics and the abstract
theory of dynamical systems in a very special domain, namely the theory of the generalized zeta-functions for dynamical systems.

Because the mathematical methods used in this work are presumably not widely known under physicists, we have collected in three
appendices the fundamentals of the relevant mathematical theories.

We scetch briefly Grothendieck's theory of nuclear operators in
Banach spaces and also the theory of positive operators in real Banach spaces. This theory goes back to Krein and Rutman respectively
in the form we are using it here to Krasnoselskii and Ladyzenskii.

Furthermore we prove an important theorem about composition operators in spaces of holomorphic functions and recall the spectral
properties of such operators. Our method will heavily rely on these
results.

We should also mention that a much more general and deeper discussion of the relations between the subjects treated in this work and
the theory of the thermodynamic formalism for certain dynamical systems
can be found in the recent book by D.Ruelle [22] .

I. THE KRAMERS-WANNIER TRANSFER MATRIX

In the first section of this chapter we introduce our notations
and some basic definitions which we will use throughout this work. We
recall furthermore in this chapter the classical Kramers-Wannier transfer matrix method for some systems with finite range interactions and
discuss an interpretation of this matrix which will allow us to generalize this method in a natural way also to systems with long range interactions. This generalization turns then out to be the Ruelle-Araki
operator .

We use the notations and definitions as given for instance in re-
ferences [22] or [51] .

I.1. Definitions and notations

Of our main concern in this work will be classical spin systems
on a one-dimensional lattice. These are systems where there is given
on each lattice site a classical spin variable. The values of this
variable range over a finite discrete or also a continuous set. In the
first case the system is called a discrete spin system, otherwise a
continuous spin system.

The spins on different lattice sites interact with each other
which can be described by an interaction potential. If a given spin
can interact only with spins a finite distance away we call the interaction of finite range otherwise we have an infinite range interaction.

It is known that such simple models describe in a rough approximation the thermodynamic behaviour of real matter like ferromagnets.
The same model serves also as an approximation for a gas or an alloy
with different kinds of atoms. Having such concrete physical systems
in mind we will first introduce the mathematical formalism for describing such lattice systems which we will use in the following discussions. The reference for this formalism is the recent book "Thermody-

namic formalism" by Ruelle $\left[22\right]$.

I.1.1.   The configuration space of lattice systems

We identify the one-dimensional lattice for convenience mathematically always with the set Z of integers.   Let then F be a discrete
finite set or some compact space which we think to be a subspace of
the finite dimensional vector space $\mathbb{R}^N$.   If F is finite and discrete
we write for the elements of F in general $F = \left\{\sigma_1,.., \sigma_d\right\}$.   In case
of a compact F we denote the elements by $\vec{x},\vec{y}$ and so on.   An example
for such a compact space F is for instance the real N-sphere S_N,
N = 1,2,.. .

The physical interpretation of this set F is obviously as the set
of possible values of the classical spin variable σ which can be a
scalar or a vector in the vector space $\mathbb{R}^N$ depending on the model
we are using.   Simple examples are the well known Ising model $\left[36\right]$ with
spin 1/2 which is described by the set F = $\left\{1/2,-1/2\right\}$, or the N-vector
Potts model $\left[113\right]$ where F is the set of all N-th roots of unity in the
complex plane $\mathbb{C}$ which we identify with the real vector space $\mathbb{R}^2$.

An example where F is a compact subset of $\mathbb{R}^N$ is the N-vector
model $\left[56\right]$.   There we have F = S_{N-1} , the N-1-sphere.

Let Λ be a finite subset of the lattice Z.   We define then the
configuration space for this subset Ω_Λ .   A configuration is given as
the set of all spin values which the spins on the lattice sites in Λ
have at a certain moment.   Mathematically the set of all configurations
can be described as $\Omega_\Lambda = \prod_{i \in \Lambda} F$, which is the direct product of $|\Lambda|$
copies of the set F where $|\Lambda|$ denotes the cardinality of Λ .   This
set Ω_Λ can be interpreted also as the set of all mappings of Λ into
the set F and a configuration is just a mapping of Λ into the set F.

Provided F is equipped with a suitable topology so that it is a com-
pact topological space also the space Ω_Λ of configurations on Λ be-

comes a compact space in the so called product topology [57]. In case
F is a finite set one can take the discrete topology where all subsets
of F are open. Introducing a topology in the space Ω_Λ allows us to
say when different configurations are near to each other and also
to consider continuous mappings of the space Ω_Λ into other topological spaces. We will come back to this immediately.

We denote the elements of the space Ω_Λ by the symbol ξ_Λ. Then
we have obviously also the representation

$$\xi_\Lambda = (\xi_{i_1}, \ldots, \xi_{i_{|\Lambda|}}) \quad \text{with } \xi_{i_k} \in F \quad \text{if } \Lambda = (i_1, \ldots, i_{|\Lambda|}) \subset Z .$$

We want to include in our description also the case where not
all possible configurations on the set Λ as defined above are really
allowed configurations from the physical point of view. If one inter-
prets for instance the elements of F as a collection of d different
kinds of atoms which make up some alloy it happens very often in nature that on neighbouring lattice sites not all possible combinations
of atoms are really allowed, in the sense that for instance atom A
cannot be next to atom B and so on. We can take this situation into
account by choosing for any $\Lambda \subset Z$ a subset $\tilde{\Omega}_\Lambda$ in the space Ω_Λ defined above and calling the elements of this set the allowed confi-
gurations for the finite subset Λ in Z.

Doing so we can then define the configuration space for the infinitely extended lattice system as

$$\Omega = \left\{ \xi \in F^Z : \text{ for all finite subsets } \Lambda \text{ in } Z \quad \xi_{|\Lambda} \in \tilde{\Omega}_\Lambda \right\}. \tag{I.1}$$

Here we denoted by F^Z the compact topological space $F^Z := \prod_{i \in Z} F$.
A configuration $\xi \in F^Z$ is therefore given as $\xi = (\xi_i)_{i \in Z}$ with $\xi_i \in F$
for all $i \in Z$. The restricted configuration $\xi_{|\Lambda}$ on $\Lambda \subset Z$, $\Lambda = (i_1, \ldots, i_{|\Lambda|})$
is defined as $\xi_{|\Lambda} := (\xi_{i_k})_{1 \leq k \leq |\Lambda|}$.

In words we can therefore say that a configuration ξ on Z is allowed exactly when all restrictions of this configuration to finite regions Λ in Z are allowed configurations for these finite regions.

The compact space Ω can easily be made a metric space by introducing the following metric $d(.,.)$ [58] :

let $\xi, \eta \in \Omega$, $\xi = (\xi_i)_{i \in Z}$, $\eta = (\eta_i)_{i \in Z}$ and let λ be a fixed positive number with $0 < \lambda < 1$. Then one defines

$$d(\xi, \eta) := \lambda^n , \qquad (I.2)$$

where $n = \inf \{|i| : \xi_i \neq \eta_i \}$.

On the space Z one has the natural action of the so called translation operator $\tau : Z \longrightarrow Z$:

$$\tau(i) := i + 1 . \qquad (I.3)$$

This mapping induces also a mapping $\widetilde{\tau} : \Omega \longrightarrow \Omega$ by

$$(\widetilde{\tau} \xi)_i := \xi_{i+1} , \qquad (I.4)$$

that means $\widetilde{\tau}$ shifts the whole configuration ξ on the lattice Z one lattice site to the left. We assumed thereby tacitly that the so defined new configuration $\widetilde{\tau} \xi$ is again an allowed configuration in the sense explained above. This in fact can be easily achieved by a suitable choice for the sets $\widetilde{\Omega}_\Lambda$. It is then fairly easy to show that the mapping $\widetilde{\tau}$ in (I.4) defines a homeomorphism with respect to the topology on the space Ω defined by the metric $d(.,.)$ in (I.2). This means both the mapping $\widetilde{\tau}$ and its inverse $\widetilde{\tau}^{-1}$ are then continuous mappings. For simplicity we will write from now on for the mapping $\widetilde{\tau}$ also τ as long as there is no danger of confusion . The operator τ is also called the shift operator on the configuration space Ω.

An important role in the physical description of a system is played
by the set of observables of such a system. This is simply the set of
all quantities which can be measured and observed for the system. In
the case of a classical system the space of observables is given mathematically by the space $\mathcal{C}(\Omega)$ of all continuous and real valued functions on the configuration space Ω [59]. We see at this point why
it is important to have a topology on this space. The space $\mathcal{C}(\Omega)$
when equipped with the sup-norm

$$\| f \| := \sup_{\xi \in \Omega} \left| f(\xi) \right| , \quad \text{for } f \in \mathcal{C}(\Omega) , \tag{I.5}$$

becomes a real Banach space. In analogy we can also define the space
of observables for any finite subset Λ in Z . This is the space
$\mathcal{C}(\Omega_\Lambda)$ of all continuous real valued functions on the space Ω_Λ .
Physically such an observable from $\mathcal{C}(\Omega_\Lambda)$ corresponds to an observable
of the infinitely extended system which can be measured in the finite
region Λ :

   If we denote by $\tilde{\alpha}_\Lambda : \Omega \longrightarrow \Omega_\Lambda$ the restriction mapping

$$\tilde{\alpha}_\Lambda(\xi) := \xi|_\Lambda , \tag{I.6}$$

we get a natural embedding $\alpha_\Lambda : \mathcal{C}(\Omega_\Lambda) \rightarrow \mathcal{C}(\Omega)$ defined as

$$\alpha_\Lambda(f_\Lambda) := f_\Lambda \circ \tilde{\alpha}_\Lambda . \tag{I.7}$$

Let $\alpha_\Lambda \mathcal{C}(\Omega_\Lambda) := \mathcal{C}_\Lambda \subset \mathcal{C}(\Omega)$. Then $\mathcal{C}_\Lambda$ is just the space of all
observables for the system on Z which can be observed and measured in
the finite region Λ .

   Another notion which is of fundamental importance for the description of any physical system is that of a state of the system.

In the case of our classical systems such a state is given quite generally by a Borel probability measure on the metric space Ω [59].

Because Ω is compact and metrizable it follows from the Riesz representation theorem [60] that every such measure is uniquely given by a continuous linear functional α_μ on the space $\mathcal{C}(\Omega)$ with $\alpha_\mu(1) = 1$ and $\alpha_\mu(f) \geq 0$ for all $f \in \mathcal{C}(\Omega)$ with $f \geq 0$.

This means that for such spaces the positive normalized functionals on the space $\mathcal{C}(\Omega)$ are identical to the probability measures μ on Ω. A positive normalized linear functional α_μ defines on the other hand just a state of our system where the expectation value of an observable $f \in \mathcal{C}(\Omega)$ is given as

$$\langle f \rangle_{\alpha_\mu} = \alpha_\mu(f) = \int_\Omega d\mu \; f \; . \tag{I.8}$$

Thereby we used in the last relation the Riesz representation theorem. In the following we will therefore identify the probability measures with the states of our system.

I.1.2. Interactions for lattice systems

We will next explain the notion of an interaction for a lattice system. Quite generally we define as an interaction any real valued continuous function Φ on the space $\bigcup_{\Lambda \subset Z} \tilde{\Omega}_\Lambda$ of all configuration spaces over the finite subsets Λ of the lattice Z which fulfills the following conditions [22] :

1) $\Phi(\tilde{\Omega}_\phi) = 0$, where ϕ denotes the empty set in Z , $\qquad$ (I.9)

2) for all $i \in Z$ the quantity $\|\Phi\|_i := \sum_{\Lambda, \Lambda \ni i} 1/|\Lambda| \sup_{\xi_\Lambda \in \tilde{\Omega}_\Lambda} |\Phi(\xi_\Lambda)|$

$\qquad$ is finite. $\qquad\qquad$ (I.10)

This last condition just means that the function Φ should decrease fast enough with the diameter of the sets Λ .

Given then an interaction Φ one can consider the energy function $U_\Lambda^\Phi : \tilde{\Omega}_\Lambda \to \mathbb{R}$ which expresses the energy contained in a specific configuration ξ_Λ in the finite region Λ . It is defined as

$$U_\Lambda^\Phi (\xi_\Lambda) : = \sum_{M \subset \Lambda} \Phi (\xi_\Lambda |_M) , \qquad (I.11)$$

where the summation runs over all finite subsets of the set Λ .

The following estimate can easily be established with the help of definition (I.10) :

$$\| U_\Lambda^\Phi \| : = \sup_{\xi_\Lambda \in \Omega_\Lambda} | U_\Lambda^\Phi (\xi_\Lambda) | = \sum_{i \in \Lambda} \| \Phi \|_i < \infty . \qquad (I.12)$$

Very important for the following discussions is also the notion of the interaction energy $W_{\Lambda, M}$ of a configuration ξ in the two regions Λ and M in Z with $\Lambda \cap M = \phi$. For Λ a finite subset of the lattice Z we define this interaction energy for a configuration $\xi \in \tilde{\Omega}_{\Lambda \cup M}$ as

$$W_{\Lambda, M}(\xi) : = \sum_X \Phi (\xi |_X) \qquad (I.13)$$

where the summation runs over all finite subsets $X \subset Z$ with $X \cap \Lambda \neq \phi$ and $X \cap M \neq \phi$.

I.1.3. The Gibbs states

Let Λ be again a finite region in Z and Φ an interaction as introduced in the last section. The Gibbs ensemble for the region Λ with interaction Φ is defined as a probability measure μ_Λ on the space Ω_Λ with

$$\mu_\Lambda \{ \xi_\Lambda \} : = Z_\Lambda^{-1} \exp - \beta U_\Lambda^\Phi (\xi_\Lambda) , \qquad (I.14)$$

where $\{\xi_\Lambda\}$ denotes the subset of the space Ω_Λ which consists of the point ξ_Λ. The quantity β in relation (I.14) is the well known Boltzmann factor $\beta = 1/kT$. $U_\Lambda^\Phi (\xi_\Lambda)$ denotes the energy of the configuration ξ_Λ and was defined in (I.11). The quantity Z_Λ finally is the partition function for the classical spin system with interaction Φ and is defined as

$$Z_\Lambda := \sum_{\xi_\Lambda \in \tilde{\Omega}_\Lambda} \exp -\beta\, U_\Lambda^\Phi (\xi_\Lambda) \quad . \tag{I.15}$$

In the standard literature the above ensemble is called the canonical ensemble, but we will call it in this work simply the Gibbs ensemble or the Gibbs measure. To be more precise we should better call it the Gibbs ensemble with vanishing boundary conditions because all spins outside the region Λ do not contribute to the energy function which amounts to the same as setting them equal to zero.

In analogy the quantity Z_Λ should be called the partition function of the spin system with vanishing boundary conditions. This distinction will be necessary when we consider also different boundary conditions later.

To obtain then a state for the infinitely extended system on the lattice Z one has to perform the so called thermodynamic limit: one considers regions Λ becoming larger and larger so that finally every bounded region M is contained in Λ. One can then show [61] that there exists under such circumstances always a subsequence of finite regions Λ_n such that

$$\lim_{n\to\infty} \alpha^*_{M,\Lambda_n} \mu_{\Lambda_n} = \rho_M$$

exists for every finite region M. Furthermore, there exists a probability measure ρ on the space Ω with

$$\rho_M = \alpha^*_{M,Z}\, \rho \quad .$$

Thereby we have introduced for finite subsets M, $M \subset \Lambda \subset Z$, the restriction mappings

$$\alpha^*_{M,\Lambda} : \mathcal{C}(\Omega_\Lambda)^* \longrightarrow \mathcal{C}(\Omega_M)^*$$ (I.16)

defined as

$$(\alpha^*_{M,\Lambda} \mu_\Lambda)(f_M) := \mu_\Lambda(\alpha_{M,\Lambda} f_M) ,$$ (I.17)

for $\mu_\Lambda \in \mathcal{C}(\Omega_\Lambda)^*$ and $f_M \in \mathcal{C}(\Omega_M)$.

The mapping $\alpha_{M,\Lambda} : \mathcal{C}(\Omega_M) \longrightarrow \mathcal{C}(\Omega_\Lambda)$ is the natural generalization of the mapping α_Λ defined in (I.7) :

$$(\alpha_{M,\Lambda} f_M)(\xi_\Lambda) := f_M(\xi_\Lambda|_M) .$$

The probability measure ϱ on the space Ω is called the thermodynamic limit of the probability measures μ_{Λ_n}.

Now we are prepared to explain the notion of a Gibbs state for the infinitely extended system on the lattice Z. These states have been introduced by Dobrushin [62], [63] and independently also by Ruelle and Lanford [64]. These authors saw the fundamental importance of these states for the whole equilibrium theory of the statistical mechanics of many particle systems. They are defined as follows:

A probability measure μ on the space Ω is called a Gibbs state for the interaction Φ if there exists for any region Λ in Z, Λ finite, a probability measure $\nu_{Z \backslash \Lambda}$ on the space $\Omega_{Z \backslash \Lambda}$ such that we have for all configurations $\xi_\Lambda \in \Omega_\Lambda$

$$(\alpha^*_\Lambda \mu)\{\xi_\Lambda\} = \int_{\Omega_{Z \backslash \Lambda}} \nu_{Z \backslash \Lambda}(d\eta) \, \mu_{\Lambda,\eta}\{\xi_\Lambda\}$$ (I.18)

where the function $\mu_{\Lambda,\eta}\{\xi_\Lambda\}$ is defined for $\eta \in \Omega_{Z\backslash\Lambda}$ as

$$\mu_{\Lambda,\eta}\{\xi_\Lambda\} := \frac{\exp-\beta\left[U_\Lambda^\Phi(\xi_\Lambda) + W_{\Lambda,Z\backslash\Lambda}(\xi_\Lambda \cup \eta)\right]}{\sum\limits_{\eta_\Lambda \in \Omega_\Lambda} \exp-\beta\left[U_\Lambda^\Phi(\eta_\Lambda) + W_{\Lambda,Z\backslash\Lambda}(\eta_\Lambda \cup \eta)\right]} . \qquad (I.19)$$

Thereby $\xi_\Lambda \cup \eta$ denotes a configuration ξ in Ω such that $\xi|_\Lambda = \xi_\Lambda$ and $\xi|_{Z\backslash\Lambda} = \eta$.

If the configuration so defined is not allowed one has to set $W_{\Lambda,Z\backslash\Lambda}(\xi) = \infty$. The symbol $\int_{\Omega_{Z\backslash\Lambda}} \nu_{Z\backslash\Lambda}(d\eta)$ in relation (I.18) denotes the linear functional corresponding to the measure $\nu_{Z\backslash\Lambda}$ on the space $\mathcal{C}(\Omega_{Z\backslash\Lambda})$ according to the Riesz theorem.

It is not difficult to see [65] that the thermodynamic limit of Gibbs ensembles defined in (I.14) is a Gibbs state on the space Ω for the interaction Φ in the sense of the definition (I.18) and (I.19). A characterization of Gibbs states which physically is a little bit easier to understand was given by Ruelle [66] with the help of so called conditional probabilities.

Theorem I.1 (Ruelle)   A probability measure μ on the space Ω is exactly then a Gibbs state if for all finite $\Lambda \subset Z$ the conditional probability for finding the configuration ξ_Λ on Λ when given the configuration η on $Z \backslash \Lambda$ is determined by the expression $\mu_{\Lambda,\eta}\{\xi_\Lambda\}$ as defined in (I.18).

The Gibbs ensembles which were introduced already by Gibbs constitute as everybody knows together with their thermodynamic limits the basis of the modern equilibrium theory of statistical mechanics. With them it was possible to deduce macroscopic properties of matter consisting of a large number of constituents from the microscopic be-

behavior of these constituents. We can not say anything at this place about the general problem of justifying these ensembles from the dynamical laws for such classical particle systems. Rather we regard the ansatz of Gibbs as a working hypothesis which has been used very successfully for a large variety of problems in equilibrium statistical mechanics which somehow justifies its utilization. For a deeper discussion of the mathematical problems connected with a derivation of such ensembles from the principles of classical mechanics we must refer to the literature [67].

One could try now certainly to deduce all the physical properties of a system alone from a study of the above Gibbs state for the system. One can ask under which conditions on the interaction ϕ there exists exactly one such Gibbs state. If this is the case then one could say that the system can exist only in one phase, or what amounts to the same, that such a system cannot have a phase transition.

It turned out however that a phase transition in the sense of non-uniqueness of the Gibbs measure does in general not coincide with the notion of a phase transition based on the analyticity behavior of the thermodynamic potentials derived from the Gibbs ensembles for the same system [68].

In this work we will restrict our discussion to phase transitions in the conventional sense, and therefore the partition function belonging to the Gibbs ensembles will be in the center of our discussions.

It is well known that all quantities of physical interest for such a system can be derived from this partition function through the thermodynamic potentials [2]. The most interesting of these is the mean free energy F which is given in terms of the partition function as

$$-\beta F_\Lambda^\Phi = |\Lambda|^{-1} \log Z_\Lambda \ , \qquad\qquad (\text{I}.20)$$

where $\Lambda \subset \mathbb{Z}$ is a finite subset and $\beta = 1/kT$.

In this work we will use only tranlation invariant interactions Φ, that means for any finite $\Lambda \subset \mathbb{Z}$ and any $\xi_\Lambda \in \Omega_\Lambda$ the function Φ satisfies the relation

$$\Phi(\xi_\Lambda) = \Phi(\tau \, \xi_\Lambda), \tag{I.21}$$

where τ is the translation operator defined in (I.4). Thereby we have again assumed that with ξ_Λ also the configuration $\tau \, \xi_\Lambda$ is an allowed configuration for our system under consideration.

The translation invariant interactions form a real Banach space $\mathcal{O}\!\ell$ with the following norm [69]

$$\|\Phi\| := \sum_{\Lambda, 0 \in \Lambda} 1/|\Lambda| \sup_{\xi_\Lambda \in \Omega_\Lambda} |\Phi(\xi_\Lambda)| < \infty . \tag{I.22}$$

Thereby 0 denotes the zero element in $\mathbb{Z}$ and the summation is over all finite subsets Λ in $\mathbb{Z}$ which contain this point.

Gallavotti and Miracle-Sole [70] proved the existence of the thermodynamic limit of the mean free energy F_Λ^Φ as defined in (I.20) for systems with an interaction from the space $\mathcal{O}\!\ell$. Let namely $\Lambda(a)$ denote the closed interval $0 \leq i \leq a$ in $\mathbb{Z}$, then the above authors proved

<u>Theorem I.2</u> (Gallavotti, Miracle-Sole) For any $\Phi \in \mathcal{O}\!\ell$ the quantity

$$f(\beta, \Phi) := \lim_{a \to \infty} F_{\Lambda(a)}^\Phi \tag{I.23}$$

exists and defines a continuous function on the Banach space $\mathcal{O}\!\ell$.

As can be seen from this theorem , the existence of the thermodynamic limit is guaranteed for any interaction Φ which fulfills the condition (I.22), that means decreases fast enough with the diameter of the set $\wedge$ in Z . Especially are there also included all kinds of n-body interactions for any $n \epsilon \mathbb{N}$. In this work here we will restrict however our discussion completely to one- and two-body interactions. Thereby we interpret a one-body interaction as the interaction of a spin variable with an external magnetic field or, in the case of a lattice gas as the chemical potential of a particle.

We will therefore assume from now on that the function Φ has the property that

$$\Phi \, (\xi_\wedge) \, = \, 0 \quad \text{for all} \quad \xi_\wedge \epsilon \, \Omega_\wedge \, \text{with } |\wedge| > 2 \, . \tag{I.24}$$

In this case the Banach space $\mathcal{O}l$ is just the space of all potentials Φ for which

$$\|\Phi\| \, = \, \sum_{i=1}^{\infty} \, \sup_{\xi_{\wedge_i} \epsilon \Omega_{\wedge_i}} \, |\Phi(\xi_{\wedge_i})| < \infty \, , \tag{I.25}$$

where $\wedge_i$ denotes the set $\{0,i\} \subset Z$ with $|\wedge_i| \, = \, 2$.

The finite range interactions are described by functions $\Phi \epsilon \, \mathcal{O}l$ for which there exists a natural number $k \epsilon \mathbb{N}$ with

$$\Phi \, (\xi_\wedge) \, = \, 0 \quad \text{for diam } \wedge \, := \, \max_{i,j \epsilon \wedge} \, |\, i-j| \, > \, k \, .$$

These interactions form a subspace of the Banach space $\mathcal{O}l$ which , as can easily be seen, is in fact dense in the space $\mathcal{O}l$.

The real mathematical problem in equilibrium statistical mechanics consists now in calculating the free energy $f(\beta, \Phi)$ as defined in (I.23). This is such a difficult problem that in spite of tremendous efforts over many years an explicit calculation of this free energy can be done only for a very small number of interactions Φ . In fact analytic expressions are known only for some models in one and two dimensions. In two dimensions the lattice can be taken to be the space $Z_2 = Z \times Z$. The best known model in this connection is certainly the two-dimensional spin 1/2 Ising model   with nearest neighbour interaction. The set F is again given as $F = \{1/2, -1/2\}$ and the interaction Φ can be described as

$$\Phi(\xi_\Lambda) = \begin{cases} 0 & \text{for all } \xi_\Lambda \in \Omega_\Lambda \text{ with } \Lambda \subset Z_2 \text{ and diam } \Lambda > 1, \\ -h\,\xi_{\vec{i}} & \text{for all } \xi_\Lambda \in \Omega_\Lambda \text{ with } \Lambda \subset Z_2 \text{ and } \Lambda = \{\vec{i}\}, \\ -J\,\xi_{\vec{i}}\,\xi_{\vec{j}} & \text{for all } \xi_\Lambda \in \Omega_\Lambda \text{ with } \Lambda \subset Z_2 \text{ and } \Lambda = \{\vec{i},\vec{j}\} \text{ and} \end{cases} \quad (I.26)$$

$$|\vec{i} - \vec{j}| = 1 .$$

L. Onsager [71] succeeded in calculating explicitly the free energy of this system in the case of vanishing exterior magnetic field h. This was perhaps the biggest success of modern statistical mechanics: it was demonstrated for the first time that the partition function Z_Λ can describe in the thermodynamic limit a system which has non-analytic thermodynamic potentials. For finite volume Λ this partition function Z_Λ is trivially an analytic function in all parameters ce also all derivatives of the logarithm of Z_Λ must be real analytic because Z_Λ cannot vanish in the physical region of the parameters. But these derivatives are just the thermodynamic potentials.

The method Onsager used in deriving his results consisted in an extension of former attempts by Kramers and Wannier [49] and also by Montroll [50] to calculate the free energy of the two dimensional Ising model in analogy to the method employed by Ising in discussing the one-dimensional model. These authors however did not succeed in solving the problem.

This method consists in a translation of the problem into an algebraic one, namely the discussion of the eigenvalues of a certain matrix in terms of which the free energy can be simply expressed. This method is called now the transfer matrix method. From now on it will be the central theme of this work.

In the next section we will recall this method first for systems with finite range interactions. Our aim thereby will be to present an interpretation of the transfer matrix which as we will see later allows a natural and straightforward generalization also for systems with long range interactions. This way we will see that the more abstract approach by Ruelle and Araki can be recognized in a much more physical manner as this has been done up to now and appears as the natural extension of the classical Ising-Kramers-Wannier approach.

I.2. The Kramers-Wannier transfer matrix for finite range interactions

One-dimensional models with finite range interactions are not very interesting from the physical point of view because they do not show the phenomenon of phase transitions. The free energy for such systems is a real analytic function in all parameters which describe such a system $[31] - [33]$. Nevertheless also among these systems only very few are exactly solvable in the sense that one can write down an explicit analytic expression for the thermodynamic potentials. In fact, only for very short ranges r of the interaction ϕ such explicit solutions are known.

To prove however analyticity of the free energy $f(\beta, \phi)$ it is fortunately not necessary to know the exact solution. There exist several methods to prove analyticity properties of $f(\beta, \phi)$ as for instance the Lee-Yang theory in connection with correlation inequalities $[72] - [74]$ or the so called infrared bounds in connection with Osterwalder-Schrader positivity $[75]$. The last mentioned method has been applied with great success also to higher dimensional models just the last few years. For finite range interactions the simplest and at the same time most efficient method so far is however the

transfer method of Kramers and Wannier. Contrary to the above mentioned infrared bounds and correlation inequalities it allows statements
about phase transitions of any order and not only of order one which
are accompanied by a spontaneous symmetry breakdown like a sponatane-
ous magnetization in a ferromagnet. Unfortunately this method is not
so well developped for long range interactions which as we tried to
explain are the real interesting ones. This we want to improve in the
present work. We start our discussion by introducing this transfer
matrix for finite range interactions to understand better our strategy
later in case of long range interactions.

I.2.1.   Discrete spin systems on a lattice

It follows from the assumed translation invariance of the interaction Φ that we can consider instead of a system on the lattice Z
a system on the positive half axis $Z_> = \{i \in Z : i \geq 1\}$. All considerations we carried out for a system on Z can be translated word by
word for a system on $Z_>$. The only exception is the statement concerning the translation operator τ in (I.3) which is no longer an inverti-
ble mapping on configuration space $\Omega_>$ for the system on $Z_>$. Thereby we will use the convention to index all quantities for the system
on $Z_>$ with the symbol" > ". For example the space $\mathcal{Y}(\Omega_>)$ denotes the
space of observables of the system on the lattice $Z_>$ and so on.

The interaction Φ with finite range r can be written as

$$
\Phi(\xi_\Lambda) = \begin{cases} - h(\xi_i) & \text{for } \xi_\Lambda = (\xi_i) \, , \, \Lambda = \{i\} c Z_> \, , \\ -J_{|j-i|}(\xi_i, \xi_j) & \text{for } \xi_\Lambda = (\xi_i, \xi_j), \Lambda = \{i,j\} c Z_> \\ & |i - j| \leq r \\ 0 & \text{otherwise.} \end{cases} \quad (I.27)
$$

Thereby h respectively J_k are real valued functions on the space F respectively F x F .

Let $\xi > \in \Omega >$ be an allowed configuration. Let Λ_{nr} be the closed interval $1 \leqslant i \leqslant nr$ in $Z >$. The partition function Z_{nr} for this interval with periodic boundary conditions is then given as

$$Z_{nr} = \sum_{\xi_\Lambda \in \Omega_{\Lambda_{nr}}} \exp{-\beta} \left[U_{\Lambda_{nr}}^{\Phi} (\xi_\Lambda) + W_{\Lambda_{nr}, Z > \setminus \Lambda_{nr}} (\xi_\Lambda \cup \xi_{Z > \setminus \Lambda_{nr}}) \right]. \qquad (I.28)$$

Thereby the configuration $\xi_{Z > \setminus \Lambda_{nr}}$ on the set $Z > \setminus \Lambda_{nr}$ is defined as

$$\xi_{Z > \setminus \Lambda_{nr}} = (\xi_{i+nr}) i \in \mathbb{N} \quad \text{with} \quad \xi_{i+nr} = \xi_i \quad \text{for all } i \in \mathbb{N} \quad \text{if the}$$

configuration ξ_Λ on Λ_{nr} is given by $\xi_\Lambda = (\xi_i)_{1 \leqslant i \leqslant nr}$.

The quantities $U_{\Lambda_{nr}}^{\Phi}$ and $W_{\Lambda_{nr}, Z > \setminus \Lambda_{nr}}$ have been defined in (I.11) and (I.13) and denote the energy of the configuration ξ_Λ in Λ_{nr} respectively the interaction energy of this configuration with the above configuration $\xi_{Z \setminus \Lambda_{nr}}$ outside the interval Λ_{nr}. If we insert in expression (I.28) the interaction given in (I.27) we get

$$Z_{nr} = \sum_{\xi_1, \dots, \xi_{nr} \in F} \exp{\beta} \left[\sum_{i=1}^{nr} \sum_{j=i+1}^{i+r} J_{j-i}(\xi_i, \xi_j) + \right.$$

$$\left. + \sum_{i=1}^{nr} h(\xi_i) \right], \qquad (I.29)$$

where $\xi_{nr+i} = \xi_i$ for $1 \leqslant i \leqslant nr$ because of the periodic boundary conditions.

As shown by Kramers and Wannier [49] for $r = 1$ and by Rushbrooke and Ursell [31] respectively Baur and Nosanow [32] for arbitrary r the quantity Z_{nr} in (I.29) can easily be calculated by introducing the following real $d^r \times d^r$ matrix $\mathbb{L}$ indexed by the r-tuples $(\sigma_1, \dots, \sigma_r) \in F^r$:

$$\mathbb{L}_{(\sigma_1, \dots, \sigma_r), (\sigma_1', \dots, \sigma_r')} := \exp{\beta} \left[\sum_{k=1}^{r-1} \sum_{l=1}^{r-k} J_l(\sigma_k', \sigma_{k+l}') + \right.$$

$$\left. + \sum_{k=1}^{r} \sum_{l=1}^{k} J_{l+r-k}(\sigma_k', \sigma_l) + \sum_{l=1}^{r} h(\sigma_l') \right]. \qquad (I.30)$$

By using this matrix the partition function Z_{nr} can then be written as

$$Z_{nr} = \sum_{\sigma_1 \in F} \cdots \sum_{\sigma_{nr} \in F} \mathbb{L}(\sigma_1, \ldots, \sigma_r), (\sigma_{r+1}, \ldots, \sigma_{2r}) \cdot$$

(I.31)

$$\cdot \mathbb{L}(\sigma_{r+1}, \ldots, \sigma_{2r}), (\sigma_{2r+1}, \ldots, \sigma_{3r}) \cdots \mathbb{L}(\sigma_{(n-1)r}, \ldots, \sigma_{nr}), (\sigma_1, \ldots, \sigma_r).$$

But this is just

$$Z_{nr} = \text{trace } \mathbb{L}^n .$$

Because the trace of a finite dimensional matrix is given by the sum over its eigenvalues counted according to their algebraic multiplicity we get

$$Z_{nr} = \sum_{\{i\}} \lambda_i^n ,$$

(I.32)

where the λ_i's are just these eigenvalues of the matrix $\mathbb{L}$.

It follows from relation (I.30) that for real h all matrix elements of the matrix $\mathbb{L}$ are strictly positive. The theorem of Perron-Frobenius [76], [77] then shows immediately that there exists a strictly positive and simple eigenvalue λ_1 of $\mathbb{L}$ which is larger than all other eigenvalues λ of $\mathbb{L}$, $\lambda \neq \lambda_1$. Therefore we can calculate the free energy $f(\beta, \Phi)$ of our system from (I.23) to be

$$-\beta f(\beta, \Phi) = \lim_{n \to \infty} 1/nr \ \log \sum_{\{i\}} \lambda_i^n = 1/r \ \log \lambda_1 .$$

(I.33)

Known theorems about the analytic behavior of eigenvalues of matrices depending holomorphically on different parameters [53] show that λ_1 is real analytic in such parameters. The same is then certainly true also for the function $-\beta f(\beta, \Phi)$ because of relation (I.33). This

just reflects the fact that in such systems with finite range interactions there cannot exist a phase transition of any order. This is
of course known for a long time and it was indeed not for this reason
why we discussed the above example in some detail. We are more interested in the interpretation of this transfer matrix $\mathbb{L}$ which will allow
us to establish a natural relation of this matrix with the abstract
Ruelle-Araki transfer operator later.

For this we consider a fixed configuration $\xi(\sigma_1,\ldots,\sigma_r)$ in the
space $\Omega_{Z\backslash\Lambda_r}$. This configuration should be given as

$$\xi(\sigma_1,\ldots,\sigma_r) = (\xi_i)_{i\,\geq\,r+1} \quad \text{with } \xi_{r+k} = \sigma_k \quad \text{for } 1 \leq k \leq r .$$

We can then ask for the conditional probability for finding on the
interval $\Lambda_r = 1 \leq i \leq r$ in Z, the configuration $\xi_{\Lambda_r} = (\sigma_1',\ldots,\sigma_r')$
if the above configuration $\xi(\sigma_1,\ldots,\sigma_r)$ is given on the interval
$Z_{>}\backslash\Lambda_r$. This conditional probability is determined for any Gibbs state according to Theorem I.1 by the expression

$$\mu_{\Lambda_r,\,\xi(\sigma_1,\ldots,\sigma_r)}\{\xi_{\Lambda_r}\} = \frac{\exp-\beta\left[U_{\Lambda_r}^{\Phi}(\xi_{\Lambda_r}) + W_{\Lambda_r,\,Z_{>}\backslash\Lambda_r}(\xi_{\Lambda_r}\cup\xi(\sigma_1,\ldots,\sigma_r))\right]}{\sum_{\eta\in\Omega_{\Lambda_r}}\exp-\beta\left[U_{\Lambda_r}^{\Phi}(\eta) + W_{\Lambda_r,\,Z_{>}\backslash\Lambda_r}(\eta\cup\xi(\sigma_1,\ldots,\sigma_r))\right]} .$$

$$(\text{I.34})$$

In particular it follows then for the system with an interaction Φ as
in (I.27) that

$$\mu_{\Lambda_r,\,\xi(\sigma_1,\ldots,\sigma_r)}\{\xi_{\Lambda_r}\} = \frac{\mathbb{L}_{(\sigma_1,\ldots,\sigma_r),\,(\sigma_1',\ldots,\sigma_r')}}{\sum_{(\bar\sigma_1,\ldots,\bar\sigma_r)\in F^r}\mathbb{L}_{(\bar\sigma_1,\ldots,\bar\sigma_r),\,(\sigma_1',\ldots,\sigma_r')}} . \quad (\text{I.35})$$

Now this shows that the matrix element $\mathbb{L}_{(\sigma_1,\ldots,\sigma_r),\,(\sigma_1',\ldots,\sigma_r')}$ is
up to a continuous function just the conditional probability

$\mu_{\Lambda_r, \xi(\sigma_1, \ldots, \sigma_r)}$.

We can therefore also consider conditional expectation values for observables $g \in \mathcal{C}(\Omega_{\Lambda_r})$ with respect to this conditional probability $\mu_{\Lambda_r, \xi(\sigma_1, \ldots, \sigma_r)}$ and get

$$\langle g \rangle_{\mu_{\Lambda_r, \xi(\sigma_1, \ldots, \sigma_r)}} = \overline{\sum_{\eta_{\Lambda_r} \in \Omega_{\Lambda_r}}} \; g(\eta_{\Lambda_r}) \; \mu_{\Lambda_r, \xi(\sigma_1, \ldots, \sigma_r)} \{\eta_{\Lambda_r}\}. \tag{I.36}$$

The right hand side defines obviously a continuous function in the configuration $\xi(\sigma_1, \ldots, \sigma_r)$ which in fact depends only on the first r spin values $(\sigma_1, \ldots, \sigma_r)$. Therefore expression (I.36) defines a linear operator $\mathcal{M} : \mathcal{C}(\Omega_{\Lambda_r}) \longrightarrow \mathcal{C}(\Omega_{\Lambda_r})$ by setting

$$\mathcal{M} g(\xi_{\Lambda_r}) := \langle g \rangle_{\mu_{\Lambda_r, \xi}} \tag{I.37}$$

for any $\xi_> \in \Omega_>$ with $\xi_>|_{\Lambda_r} = \xi_{\Lambda_r}$.

The above operator $\mathcal{M}$ was introduced by Ledrappier [78] who used it to discuss uniqueness properties of Gibbs states for several one-dimensional lattice systems.

Closely related to the operator $\mathcal{M}$ is another operator $\mathcal{L} : \mathcal{C}(\Omega_{\Lambda_r}) \to \mathcal{C}(\Omega_{\Lambda_r})$ when one replaces the conditional measure $\mu_{\Lambda_r, \xi}(\sigma_1, \ldots, \sigma_r)$ in relation (I.36) by another Borel measure $\tilde{\mu}_{\Lambda_r, \xi}(\sigma_1, \ldots, \sigma_r)$ on the space Ω_{Λ_r} which is related to the former simply as follows

$$\tilde{\mu}_{\Lambda_r, \xi}(\sigma_1, \ldots, \sigma_r) = \mu_{\Lambda_r, \xi}(\sigma_1, \ldots, \sigma_r) \overline{\sum_{\eta_{\Lambda_r} \in \Omega_{\Lambda_r}}} \exp{-\beta} \left[U^{\Phi}_{\Lambda_r}(\eta_{\Lambda_r}) + \right.$$

$$\left. + W_{\Lambda_r, \mathbb{Z} \setminus \Lambda_r}(\eta_{\Lambda_r} \cup \xi(\sigma_1, \ldots, \sigma_r)) \right] . \tag{I.38}$$

In account of relation (I.35) we then get

$$\tilde{\mu}_{\Lambda_r, \, \xi(\sigma_1, \dots, \sigma_r)} \{\eta_{\Lambda_r}\} = \mathbb{L}_{(\sigma_1, \dots, \sigma_r), (\sigma_1', \dots, \sigma_r')} \qquad (I.39)$$

in case the configuration $\eta_{\Lambda_r} \in \Omega_{\Lambda_r}$ is given as $\eta_{\Lambda_r} = (\sigma_i')_{1 \leq i \leq r}$.

The operator $\mathscr{L} : \ell(\Omega_{\Lambda_r}) \longrightarrow \ell(\Omega_{\Lambda_r})$ is then defined as

$$\mathscr{L}_{g}(\xi_{\Lambda_r}) : = \sum_{\eta_{\Lambda_r} \in \Omega_{\Lambda_r}} \tilde{\mu}_{\Lambda_r, \, \xi(\sigma_1, \dots, \sigma_r)} \{\eta_{\Lambda_r}\} \, g(\eta_{\Lambda_r}) \qquad (I.40)$$

with $\xi_{\Lambda_r} = (\sigma_i)_{1 \leq i \leq r}$.

In words one can say that the quantity $\mathscr{L}g$ gives just the expectation of the observable $g \in \ell(\Omega_{\Lambda_r})$ with respect to the measure
$\tilde{\mu}_{\Lambda_r, \, \xi(\sigma_1, \dots, \sigma_r)}$. This value depends on the configuration $\xi(\sigma_1, \dots, \sigma_r)$
in $Z \setminus \Lambda_r$ because of the finite range of the interaction Φ only through
the first r spin values $(\sigma_1, \dots, \sigma_r)$ of this configuration. This reflects
just the Markov character of such a system with finite range interaction [79].

Using relation (I.35) we can write the operator $\mathscr{L}$ also in the
form

$$\mathscr{L}_{g}(\xi_{\Lambda_r}) = \sum_{\eta_{\Lambda_r} \in \Omega_{\Lambda_r}} \mathbb{L}_{(\sigma_1, \dots, \sigma_r), (\sigma_1', \dots, \sigma_r')} \, g(\eta_{\Lambda_r}) \qquad (I.41)$$

where the configuration η_{Λ_r} is given as $\eta_{\Lambda_r} = (\sigma_i')_{1 \leq i \leq r}$.

This now shows that the transfer matrix $\mathbb{L}$ can be interpreted also
as a linear operator in the space $\ell(\Omega_{\Lambda_r})$ of those observables of our
spin system which can be measured on the interval $\Lambda_r = [1, r]$ in $Z_>$.

This gives us an interpretation of the transfer matrix which can

be generalized in a natural way also to systems with long range interactions as we will see later.

The case of a discrete spin system with a finite range interaction
on a one-dimensional lattice discussed above is of special simplicity
because it leads to a transfer matrix L which acts in a finite dimensional vector space, in the above case just $\mathbb{R}^{d^r}$. That this is however
not always so even for lattice systems with a finite range interaction
is known from continuous spin systems like the N-vector model [56].
The N-vector model is for N = 1 identical to the classical Ising model, for N = 3 it describes the classical Heisenberg model [80]. For
the case of nearest neighbour interaction, that means r = 1, the N = 2
model was solved exactly by Joyce [80] and the N = 3 model by Rae [81]
both for vanishing exterior magnetic field. Stanley [56] discussed
the general case N with isotropic interaction.

In the next section we will consider the transfer matrix for a general continuous spin system on $Z_>$ with finite range interaction and
discuss the interpretation of this matrix analogous to the discrete
case.

I.2.2. Continuous spin systems on a lattice

Let F be any metrizable compact space with a Borel measure $d\omega$ on
it. For the N-vector model mentioned above one has for instance

$$F = S_{N-1} = \left\{ \vec{x} \in \mathbb{R}^N : \|\vec{x}\| = 1 \right\} . \tag{I.42}$$

We will then denote the elements of F quite generally by $\vec{x}$. The configuration space of such a system on the lattice $Z_>$ is again given as
$\Omega_> = \prod_{i \in Z_>} F$. We write a general configuration $\xi_> \in \Omega_>$ also as
$\xi_> = (\vec{\xi}_i)_{i \in \mathbb{N}}$ with $\vec{\xi}_i \in F$ for all $i \in \mathbb{N}$.

The interaction Φ for the continuous spin system we take as

$$\Phi(\vec{\xi}_\Lambda) = \begin{cases} -h(\vec{\xi}_i) & \text{for } \vec{\xi}_\Lambda \in \Omega_\Lambda \text{ and } \Lambda = \{i\} \subset Z_> \\ -M_{|j-i|}(\vec{\xi}_i, \vec{\xi}_j) & \text{for } \vec{\xi}_\Lambda \in \Omega_\Lambda, \Lambda = \{i,j\}, |i-j| \le r \\ 0 & \text{otherwise.} \end{cases}$$ (I.43)

Thereby the M_i's and h are real valued functions on $F \times F$ respectively F. In the case of the N-vector model they are given as $M(\vec{x}, \vec{y}) = \vec{x} \cdot \vec{y}$ and $h(x) = \vec{h} \cdot \vec{x}$ for some vector $\vec{h} \in \mathbb{R}^N$.

Let Λ_{nr} denote again the interval $1 \le i \le nr$ in $Z_>$ and take an arbitrary configuration $\vec{\xi}_{\Lambda_{nr}} \in \Omega_{\Lambda_{nr}}$. According to (I.43) we then get for the energy of this configuration

$$U_{\Lambda_{nr}}^\Phi(\vec{\xi}_{\Lambda_{nr}}) = -\sum_{i=1}^{(n-1)r} \sum_{j=1}^{r} M_j(\vec{\xi}_i, \vec{\xi}_{i+j}) + \sum_{i=(n-1)r+1}^{nr-1} \sum_{j=1}^{nr-i} M_j(\vec{\xi}_i, \vec{\xi}_{i+j})$$

$$+ \sum_{j=1}^{nr} h(\vec{\xi}_j) \ . \tag{I.44}$$

The partition function Z_{nr} with periodic boundary conditions then reads

$$Z_{nr} = \int_F d\omega_1 \cdots \int_F d\omega_{nr} \ \exp-\beta\left[U_{\Lambda_{nr}}^\Phi(\vec{\xi}_{\Lambda_{nr}}) + W_{\Lambda_{nr}, Z_> \backslash \Lambda_{nr}}(\vec{\xi}_{\Lambda_{nr}} \cup \vec{\eta}) \right]$$ (I.45)

where $\vec{\eta}$ denotes the configuration on $Z_> \backslash \Lambda_{nr}$ which one gets by perio-
dic continuation of the configuration $\vec{\xi}_{\Lambda_{nr}}$ outside the region Λ_{nr}.
That means $\vec{\eta}_{nr+i} = \vec{\eta}_i$ for all $i \in \mathbb{N}$.
Inserting expression (I.44) into the above relation we get

$$Z_{nr} = \int_F d\omega_1 \cdots \int_F d\omega_{nr} \ \exp\beta\left[\sum_{i=1}^{nr} \sum_{j=1}^{r} M_j(\vec{\xi}_i, \vec{\xi}_j) + \sum_{i=1}^{nr} h(\vec{\xi}_i) \right].$$

$$\tag{I.46}$$

To get now the transfer matrix for this system we first define a
linear operator $\mathcal{L}: \mathcal{L}_2(F \times .. \times F) \longrightarrow \mathcal{L}_2(F \times .. \times F)$ as follows

$$\mathcal{L} f(\vec{x}_1,..,\vec{x}_r) := \int_F d\omega'_1 .. \int_F d\omega'_r \ \exp\beta \left[\sum_{i=1}^{r-1} \sum_{j=1}^{r-i} M_j(\vec{x}'_i,\vec{x}'_{i+j}) + \right.$$

$$\left. + \sum_{i=1}^{r} \sum_{j=1}^{i} M_{r+j-i}(\vec{x}'_i,\vec{x}_j) + \sum_{i=1}^{r} h(\vec{x}'_i) \right] f(\vec{x}'_1,..,\vec{x}'_r) \ .$$

$$(I.47)$$

The operator $\mathcal{L}$ is therefore an integral operator in the Hilbert space
$\mathcal{L}_2(F^r)$ of square integrable functions over the r-fold product of the
space F. Its kernel $\mathcal{L}(\vec{x}_1,..,\vec{x}_r,\vec{x}'_1,..,\vec{x}'_r)$ is given by the expression

$$\mathcal{L}(\vec{x}_1,..,\vec{x}_r;\vec{x}'_1,..,\vec{x}'_r) = \exp\beta \left[\sum_{i=1}^{r-1} \sum_{j=1}^{r-i} M_j(\vec{x}'_i,\vec{x}'_{i+j}) + \sum_{i=1}^{r} h(\vec{x}'_i) + \right.$$

$$\left. + \sum_{i=1}^{r} \sum_{j=1}^{i} M_{r+j-i}(\vec{x}'_i,\vec{x}_j) \right] \ . \qquad (I.48)$$

Using this the partition function Z_{nr} in (I.46) can be written as

$$Z_{nr} = \int_F d\omega_1 .. \int_F d\omega_{rn} \ \mathcal{L}(\vec{x}_1,..,\vec{x}_r;\vec{x}_{r+1},..,\vec{x}_{2r}) \ \mathcal{L}(\vec{x}_{r+1},..,\vec{x}_{2r};$$

$$\vec{x}_{2r+1},..,\vec{x}_{3r}) .. \mathcal{L}(\vec{x}_{(n-1)r},..,\vec{x}_{nr};\vec{x}_1,..,\vec{x}_r) \ . \qquad (I.49)$$

From expression (I.48) one deduces immediately that the operator $\mathcal{L}$
defines in fact a Hilbert-Schmidt operator in the space $\mathcal{L}_2(F^r)$. If
moreover the functions M_j and h are even infinitely often differentiable in all the variables $\vec{x}_i$ and $\vec{x}'_i$ it follows immediately that the
operator $\mathcal{L}$ is in fact a nuclear operator of order zero in the Hilbert
space $\mathcal{L}_2(F^r)$ [83] . Thereby we have certainly assumed the space F to be
a smooth manifold so that the notion of differentiability of a function
on this space makes sense. But this shows that the operator $\mathcal{L}$ and

all its iterates have a trace which is given for the operator $\mathcal{L}$ by the formula

$$\text{trace}\,\mathcal{L} = \int_F d\omega_1 \cdots \int_F d\omega_r \; \mathcal{L}(\vec{x}_1,..,\vec{x}_r;\vec{x}_1,..,\vec{x}_r) \; .$$

From this we then get for expression (I.49)

$$z_{nr} = \text{trace}\,\mathcal{L}^n \; . \tag{I.50}$$

This shows that the operator $\mathcal{L}$ is nothing but the transfer matrix for the above system. To get finally from relation (I.50) an expression for the free energy $f(\beta,\phi)$ one can use a theorem of Jentzsch [84] which is just a generalization of the results of Perron and Frobenius to positive integral operators, that means to infinite dimensional matrices in a Hilbert space:

Theorem I.3 (Jentzsch) Let T be a linear integral operator in the Hilbert space $\mathcal{L}_2(K,ds)$ of square integrable functions f on the compact space K with respect to a Borel measure ds. If the kernel k(t,s) of this operator T is continuous and strictly positive that means k(t,s) > 0 for all $t,s \in K$, then there exists a positive eigenvalue λ_1 with an eigenfunction $f_1 > 0$ such that

1) λ_1 is a simple eigenvalue,
2) $\lambda_1 > |\lambda|$ for all $\lambda \in \sigma(T), \lambda \neq \lambda_1$, where $\sigma(T)$ denotes the spectrum of T which consists only of eigenvalues, because T is compact.

By applying this theorem we get for the free energy $f(\beta,\phi)$ the expression

$$-\beta\,f(\beta,\phi) = 1/r \; \log \lambda_1 \tag{I.51}$$

and therefore the free energy is again a real analytic function in all relevant parameters.

As for the discrete spin system we can regard the transfer matrix also for this system as a linear operator in the space $\mathcal{L}(\Omega_{\Lambda_r})$ of observables of the system which can be measured and observed in the finite interval Λ_{rn}. The difference when compared to the former case of a discrete system is only the fact that in the present case it is no longer a finite dimensional matrix but instead a trace class operator in an infinite dimensional Hilbert space. This comes from the fact that the space of observables of a continuous spin system on any finite interval is already an infinite dimensional space. The spectral properties however are in both cases completely analogous and follow indeed from the positivity properties of the matrix elements and the kernel of the integral operator. We will see later that such positivity properties are of great importance also in the discussion of the transfer matrix for long range interactions.

We mention here only for the sake of completeness that completely analogous considerations as carried out here for one-dimensional lattice systems can be applied also in higher dimensions. For a precise discussion of these higher dimensional systems we refer however to the literature.

In analogy to one-dimensional lattice systems transfer matrices can be found also for continuous hard rod systems on the real line. These are systems where extended objects move under the influence of a mutual interaction on $\mathbb{R}$. Such systems have been studied by Tonks [86], Takahasi [87] and for arbitrary finite range of the interaction by van Hove [33]. He showed that the partition function of such a system can be described in the so called pressure ensemble by an integral operator. Its spectrum can again be characterized by Theorem I.3 of Jentzsch. Because we will use in this work only canonical and grand canonical Gibbs ensembles we do not discuss the method of van

Hove in detail here. We will come back however in the next chapter
to this continuous hard rod system with finite range interaction in
connection with our discussion of the Ruelle-Araki transfer operator
and its relation to the Kramers-Wannier transfer matrix for such systems with finite range interactions. We will see that there can be
defined a transfer matrix for the continuous hard rod system also in
the grand canonical ensemble.

At this point we can summarize our considerations for one-dimensional systems with finite range interactions. These systems have a
transfer matrix which for discrete spin systems is a linear operator
in a finite dimensional vector space. For continuous systems it is
a trace class operator in an infinite dimensional Hilbert space.
The interpretation of this operator which will be of importance
for our future considerations is as follows:
    the transfer matrix defines a linear operator in the space of ob-
servables of the infinitely extended system which can be measured
in certain finite regions $\wedge(r) \subset \mathbb{Z}$ respectively in $\mathbb{R}$. This region
depends of course on the range r of the interaction. The spaces of
observables are in general infinite dimensional Banach spaces which
reduce to finite dimensional spaces for observables which can be mea -
sured in finite regions as long as the system is discrete.
    The transfer matrix applied to such an observable gives just cer-
tain conditional expectation values of these observables.

We will see in the next chapter how this interpretation of the
transfer matrix can be extended in a natural way also to systems with
long range interactions via the abstract Ruelle-Araki transfer operator which appears this way only as a generalization of the Kramers-
Wannier matrix which we discussed in some detail in this chapter.

II. THE RUELLE-ARAKI TRANSFER OPERATOR FOR ONE-DIMENSIONAL CLASSICAL SYSTEMS

II.1. General properties

We start again with lattice systems. Let $\Omega = F^Z$ be the configuration space where F is some discrete or compact set. We assume the interactions translation invariant which allows us to restrict the discussion immediately to a system on the positive half axis $Z_>$.

Let μ be a Gibbs measure on $\Omega_>$ characterized by Theorem I.1 . Let furthermore $\mu_{\Lambda,\eta}$ denote the conditional probabilities as introduced in (I.19) and $\mathcal{C}(\Omega_>)$ the space of observables of our system on the half axis $Z_>$. For Λ some finite interval in $Z_>$ we define an operator $\mathcal{M}_\Lambda : \mathcal{C}(\Omega_>) \rightarrow \mathcal{C}(\Omega_>)$ as

$$\mathcal{M}_\Lambda f(\xi_>) := \sum_{\xi_\Lambda \in \Omega_\Lambda} \mu_{\Lambda,\eta} \{\xi_\Lambda\} \, f(\xi_\Lambda \cup \xi_>) . \qquad (II.1)$$

From expression (I.37) we see that the operator $\mathcal{M}_\Lambda$ is just an extension of the operator $\mathcal{M}$ to the whole space $\mathcal{C}(\Omega_>)$. The operator $\mathcal{M}$ applied to some function f describes, as we have seen before certain conditional expectation values of the observable f on subsets of $Z_>$. The same is then true also for the operator $\mathcal{M}_\Lambda$ on the whole space of observables $\mathcal{C}(\Omega_>)$ which can therefore be interpreted in exactly the same way from a physical point of view.

The importance of the operator $\mathcal{M}_\Lambda$ for discussing the physical properties of our system follows from the next theorem [78]:

<u>Theorem II.1</u> (Ledrappier) There exists exactly one Gibbs state μ for the lattice system (Z, F, Φ) if, and only if for all $f \in \mathcal{C}(\Omega_>)$ the limit $\lim_{\Lambda \to \infty} \mathcal{M}_\Lambda f$ = constant uniformly in $\Omega_>$.

Without being able to enter the exact proof one sees already from the definition of the operator $\mathcal{M}_\Lambda$ that the quantity $\mathcal{M}_\Lambda f$ converges in the limit $\Lambda \to \infty$ to an expectation value of the observable f. If this expectation value does not depend on the way how Λ tends to infinity there exists exactly one state which gives just this expectation value for the observable f.

From the Ledrappier operator $\mathcal{M}_\Lambda$ there is not a long way to the Ruelle-Araki operator. We consider instead of the probability measure $\mu_{\Lambda,\eta}$ the measure $\tilde{\mu}_{\Lambda,\eta}$ as defined in (I.38)

$$\tilde{\mu}_{\Lambda,\eta} : = \mu_{\Lambda,\eta} \sum_{\eta_\Lambda \in \Omega_\Lambda} \exp{-\beta} \left[U_\Lambda^\Phi (\eta_\Lambda) + W_{\Lambda,Z_> \backslash \Lambda} (\eta_\Lambda \cup \eta) \right], \qquad (II.2)$$

where $\Lambda \subset Z_>$ is any finite interval and $\eta \in \Omega_{Z_> \backslash \Lambda}$ denotes some configuration outside this interval Λ.

This positive Borel measure defines a linear operator $\mathcal{L}_\Lambda$ in the space $\ell(\Omega_>)$ which is defined in analogy to the operator $\mathcal{M}_\Lambda$ in (II.1):

$$\mathcal{L}_\Lambda f(\xi_>) : = \sum_{\eta_\Lambda \in \Omega_\Lambda} \tilde{\mu}_{\Lambda,\xi_>} \{\eta_\Lambda\} \, f(\eta_\Lambda \cup \xi_>) . \qquad (II.3)$$

Inserting into this definition the explicit form of the measure $\tilde{\mu}_{\Lambda,\xi_>}$ we get

$$\mathcal{L}_\Lambda f(\xi_>) = \sum_{\eta_\Lambda \in \Omega_\Lambda} \exp{-\beta} \left[U_\Lambda^\Phi (\eta_\Lambda) + W_{\Lambda,Z_>\backslash\Lambda} (\eta_\Lambda \cup \xi_>) \right] f(\eta_\Lambda \cup \xi_>) . \qquad (II.4)$$

If we compare this expression with relation (I.40) we see that the operator $\mathcal{L}_\Lambda$ can be considered as the extension of the operator $\mathcal{L}$ to the whole space of observables of the infinitely extended system on $Z_>$. Therefore the quantity $\mathcal{L}_\Lambda f$ can be regarded again as a conditional expectation value of the observable f.

The operator $\mathcal{L}_\Lambda$ was the first time introduced by Araki [46] when he discussed one-dimensional quantum spin systems on a lattice. Later Ruelle [51] and Gallavotti respectively Miracle-Sole [52] emphasized the importance of this operator also for classical spin systems and continuous hard rod systems in one dimension.

For $\Lambda = [1,r] \subset Z_>$, the operator $\mathcal{L}_\Lambda$ can be written also as follows as one can see immediately

$$\mathcal{L}_\Lambda = \mathcal{L}_{\Lambda(1)} \circ \cdots \circ \mathcal{L}_{\Lambda(r)} \quad ,$$

where $\Lambda(i)$ denotes the one point set $\{i\}$ in $Z_>$. Consequently it is enough to consider the operator $\mathcal{L} : \mathcal{C}(\Omega_>) \longrightarrow \mathcal{C}(\Omega_>)$ defined as

$$\mathcal{L} f(\xi_>) : = \sum_{\sigma \in F} \exp-\beta \left[U_{\{1\}}^{\Phi}(\sigma) + W_{\{1\}, Z_> \backslash \{1\}}(\sigma \cup \xi_>) \right] f(\sigma \cup \xi_>).$$

(II.5)

This operator will be from now on the main object of our study and we will call it the Ruelle-Araki operator or sometimes also the Ruelle-Araki transfer matrix.

Before discussing the exact relation between this operator and the transfer matrices discussed in the previous chapter we first recall without proof a theorem of Ruelle which in fact shows the importance of this operator for such one-dimensional spin systems on a lattice. For this we define a suitable class of interactions which decrease in a certain sense fast enough at infinity. Let Φ be a translation invariant two body interaction such that

$$\||\Phi\|| : = \sum_{i=1}^{\infty} i \sup_{\xi_{\Lambda_i} \in \Omega_{\Lambda_i}} |\Phi(\xi_{\Lambda_i})| < \infty \quad , \tag{II.6}$$

where Λ_i denotes the interval $[0,i]$ in Z.

All Φ's with $\||\Phi\|| < \infty$ form a real Banach space $\mathcal{B}_1$

$$\mathcal{B}_1 : = \{ \Phi : \||\Phi\|| < \infty \} . \tag{II.7}$$

Then Ruelle proved [51]

Theorem II.2 (Ruelle)   Let $\Phi \in \mathcal{B}_1$. Then there exist exactly one
positive number $\lambda_1 > 0$, a probability measure γ on configuration space $\Omega_>$ and a positive function u in $\mathscr{C}(\Omega_>)$ with the properties

1)   u is an eigenfunction of the operator $\mathcal{L}$ with eigenvalue λ_1 :
      $\mathcal{L}u = \lambda_1 u$; γ is an eigenvector to the same eigenvalue λ_1 of the
      dual operator $\mathcal{L}^*$ of $\mathcal{L}$ in the space $\mathscr{C}(\Omega_>)^*$: $\mathcal{L}^*\gamma = \lambda_1 \gamma$ and
      $\gamma(u) = 1$.

2)   For any observable $g \in \mathscr{C}(\Omega_>)$, $g \neq 0$ we have
      $$\lim_{n \to \infty} \| \lambda_1^{-n} \mathcal{L}^n g - \gamma(g) u \| = 0 .$$

3)   The free energy $f(\beta, \Phi)$ of the system on the lattice Z is given
      as $-\beta f(\beta, \Phi) = \log \lambda_1$.

4)   The eigenvalue λ_1 is continuously differentiable in the interaction
      Φ in every finite dimensional subspace of the Banach space $\mathcal{B}_1$.

5)   The eigenvector u can be written as

      $$u(\xi_>) = c \int_{\Omega_<} \mu_<(d\xi_<) \; \exp{-\beta \; W_{z_<, z_>}(\xi_< \cup \xi_>)}$$

      where $\mu_<$ denotes the Gibbs measure for the system with interaction
      Φ on the negative half axis $Z_<$ and c is some constant.

Property 4) in the above theorem shows in particular that such spin
systems with an interaction $\Phi \in \mathcal{B}_1$ do not have a phase transition of
first order. Phase transitions of higher order however are not exclu-
ded by this theorem because the highest eigenvalue λ_1 is not ne-
cessarily an analytic function in the interaction Φ. To prove analyticity of the highest eigenvalue the above theorem has to be sharp-
end quite a lot which in fact has been done only for exponentially
decreasing interactions up to now. We will come back to this question
in detail in the third chapter.

Next we want to discuss a little bit more in detail the relation between the operator $\mathcal{L}$ in (II.5) and the Kramers-Wannier transfer matrix which we considered in the previous chapter. It is clear from Theorem II.2 that there must exist such a connection because both the Ruelle-Araki operator and the Kramers-Wannier matrix determine the free energy of a system with short range interaction via their highest eigenvalue.

II.2. On the relation between the Ruelle-Araki operator and the Kramers-Wannier matrix

II.2.1. Discrete spin systems with finite range interaction

We consider discrete spin systems on a one-dimensional lattice with an interaction Φ of the form given in (I.27). For such an interaction the Ruelle-Araki operator $\mathcal{L}$ in the space $\ell(\Omega_{>})$ looks like

$$\mathcal{L} f(\xi_{>}) = \sum_{i=1}^{d} f(\sigma_i \cup \xi_{>}) \exp\beta\left[h(\sigma_i) + \sum_{k=1}^{r} J_k(\sigma_i, \xi_k)\right], \quad (II.8)$$

where we denoted the elements of the set F by $\sigma_1, \ldots, \sigma_d$ and where the configuration $\xi_{>} \in \Omega_{>}$ is given as $\xi_{>} = (\xi_k)_{k \in \mathbb{N}}$, $\xi_k \in F$ for all $k \in \mathbb{N}$.

We see from definition (II.8) that the operator $\mathcal{L}$ leaves a certain subspace $\ell_r \subset \ell(\Omega_{>})$ invariant. In fact let ℓ_r be the space of all observables f which depend only on the first r spin values $(\xi_1, \ldots, \xi_r)$ of a configuration $\xi = (\xi_i)_{i \in \mathbb{N}}$ and which can therefore be measured in the interval $1 \leq i \leq r$ of the half axis $Z_{>}$. It is obvious that these observables form a subspace of the space $\ell(\Omega_{>})$. For such a f we get from (II.8) :

$$\mathcal{L} f(\xi_{>}) = \sum_{i=1}^{d} f(\sigma_i, \xi_1, \ldots, \xi_{r-1}) \exp\beta\left[h(\sigma_i) + \sum_{k=1}^{r} J_k(\sigma_i, \xi_k)\right]. \quad (II.9)$$

But this shows that the function $\mathcal{L}f$ belongs again to the space of observables depending only on the first r spin values of a configuration $\underline{\sigma}_{>} = (\sigma_i)_{i \in \mathbb{N}}$. In particular it is obvious from the representation for the eigenvector u belonging to the highest eigenvalue λ_1 as given in Theorem II.2 that this function belongs to the space $\mathcal{l}_r$.

The continuous functions over a discrete space with d^r elements form a finite dimensional vector space of dimension d^r. But this shows that the space $\mathcal{l}_r$ is isomorphic to the real vector space $\mathbb{R}^{d^r}$.

Every linear operator in a finite dimensional space can be represented by a matrix. In our case there exists therefore a real $d^r \times d^r$ matrix $\mathbb{L}$ which gives a representation of the operator $\mathcal{L}$ in the space $\mathcal{l}_r$. Its matrix elements $\mathbb{L}_{(\sigma_1,\ldots,\sigma_r),(\sigma'_1,\ldots,\sigma'_r)}$ can be read off immediately from relation (II.9):

Since

$$\mathcal{L}f(\sigma_1,\ldots,\sigma_r) = \sum_{(\sigma'_1,\ldots,\sigma'_r) \in F^r} \mathbb{L}_{(\sigma_1,\ldots,\sigma_r),(\sigma'_1,\ldots,\sigma'_r)} f(\sigma'_1,\ldots,\sigma'_r)$$

we get

$$\mathbb{L}_{(\sigma_1,\ldots,\sigma_r),(\sigma'_1,\ldots,\sigma'_r)} = \delta_{\sigma'_2,\sigma_1} \cdots \delta_{\sigma'_r,\sigma_{r-1}} \exp\beta\left[h(\sigma'_1) + \sum_{k=1}^{r} J_k(\sigma'_1, k)\right]. \qquad (II.10)$$

To make the relation of this matrix with the transfer matrix of the last chapter more transparent we calculate the matrix elements of the matrix $\mathbb{L}^r$. If we denote these matrix elements by $\mathbb{L}^r_{(\sigma_1,\ldots,\sigma_r),(\sigma^{(r)}_1,\ldots\sigma^{(r)}_r)}$ we get after a simple calculation which we omit here

$$\mathbb{L}^r_{(\sigma_1,\ldots,\sigma_r),(\sigma^{(r)}_1,\ldots\sigma^{(r)}_r)} = \sum_{\underline{\sigma}^{(1)},\ldots,\underline{\sigma}^{(r)}} \left(\prod_{s=1}^{r} \prod_{k=1}^{r} \delta_{\sigma^{(s)}_k,\sigma^{(s-1)}_{k-1}}\right) \cdot$$

$$\exp \beta \left[\sum_{1=1}^{r} h(\sigma_1^{(1)}) + \sum_{1=1}^{r} \sum_{k=1}^{r} J_k(\sigma_1^{(1)}, \sigma_k^{(1-1)}) \right]. \qquad (II.11)$$

Thereby we used the abbreviation

$$\underline{\sigma}^{(i)} = (\sigma_1^{(i)}, \ldots, \sigma_r^{(i)}) \quad \text{for } 1 \le i \le r.$$

The summation in (II.11) extends over all configurations $\underline{\sigma}^{(i)} \in \Omega_{\Lambda_r}$.
Using simplest properties of the Kronecker symbols like

$$\sum_{\sigma' \in F} \delta_{\sigma_k, \sigma'} \delta_{\sigma', \sigma_n} = \delta_{\sigma_k, \sigma_n}$$

the summation in (II.11) can easily be carried out.
By using furthermore the identity

$$\prod_{s=1}^{r} \prod_{k=1}^{r} \delta_{\sigma_k^{(s)}, \sigma_{k-1}^{(s-1)}} = \prod_{s=1}^{r-1} \prod_{k=s+1}^{r} \delta_{\sigma_k^{(s)}, \sigma_{k-s}^{(1)}} \prod_{s=1}^{r-1} \prod_{k=1}^{s} \delta_{\sigma_k^{(s)}, \sigma_{k+r-s}^{(r)}}$$

$$(II.12)$$

one finally can write expression (II.11) as follows

$$\mathbb{L}^r_{(\sigma_1, \ldots, \sigma_r), (\sigma_1^{(r)}, \ldots, \sigma_r^{(r)})} = \exp \beta \left[\sum_{1=1}^{r} h(\sigma_{r-1+1}^{(r)}) + \sum_{1=1}^{r} \sum_{k=1}^{r} \cdot \right.$$

$$\left. J_k(\sigma_{r-1+1}^{(r)}, \sigma_{k-1+1}^{(r)}) + \sum_{1=2}^{r} \sum_{k=1}^{1-1} J_k(\sigma_{r-1+1}^{(r)}, \sigma_{k+r+1-1}^{(r)}) \right]$$

After reordering the summation we arrive then at the following result
for the matrix elements $\mathbb{L}^r_{(\sigma_1, \ldots, \sigma_r), (\sigma_1^{(r)}, \ldots, \sigma_r^{(r)})}$:

$$\mathbb{L}^r_{(\sigma_1, \ldots, \sigma_r), (\sigma_1^{(r)}, \ldots, \sigma_r^{(r)})} = \exp \beta \left[\sum_{1=1}^{r} h(\sigma_1^{(r)}) + \sum_{1=1}^{r} \sum_{k=1}^{1} J_{r+k-1}(\sigma_1^{(r)}, \sigma_k) + \right.$$

$$\left. + \sum_{1=1}^{r-1} \sum_{k=1+1}^{r} J_{k-1}(\sigma_1^{(r)}, \sigma_k^{(r)}) \right]. \qquad (II.13)$$

We can compare this result with expression (I.30) for the matrix elements of the transfer matrix à la Kramers-Wannier for the same system with an interactions of finite range r and see that they are identi - cal.

We conclude therefore that the Ruelle-Araki operator $\mathcal{L}$ for a discrete spin system on a lattice with finite range interaction gives exactly the classical Kramers-Wamnier transfer matrix for this system if its domain of definition is restricted appropriately to a certain subspace of the Banach space $\mathcal{C}(\Omega_{>})$. To be more exact this restriction leads to the r-th root of the transfer matrix. This shows already that the Ruelle-Araki operator is without any doubt a natural generalization of this old concept of a transfer matrix **for** such systems.

Next we will show that exactly the same is true also for continuous spin systems on a one-dimensional lattice.

II.2.2. Continuous spin systems with finite range interactions

For this purpose we consider a general compact space F with some Borel measure $d\omega$ on it. It is even not necessary for F to be compact for the following discussion to be carried out. One can take any topological space F with a finite measure $d\omega$ if only the existence of the thermodynamic limit for such a system can be shown [88].
Here we restrict ourselves to compact spaces F. The Ruelle-Araki operator $\mathcal{L}$ has now the form

$$\mathcal{L} \, f(\xi_{>}) = \int_{F} d\omega \; f(\vec{x}, \xi_{>}) \; \exp-\beta\left[U^{\Phi}_{\{1\}}(\vec{x}) + W_{\{1\}, Z_{>}\setminus\{1\}}(\vec{x} \cup \xi_{>})\right] \quad , \quad (II.14)$$

where we used the same notation as in relations (I.11) and (I.13). This expression reads in the case of an interaction with finite range r as given in (I.43)

$$\mathcal{L}f(\xi_{>}) = \int_{F} d\omega \; f(\vec{x}, \xi_{>}) \; \exp-\beta\left[h(\vec{x}) + \sum_{i=1}^{r} M_{i}(\vec{x}, \vec{\zeta}_{i})\right] , \qquad (II.15)$$

where the configuration $\xi_> \in \Omega_>$ is given as $\xi_> = (\vec{\xi}_i)_{i \in \mathbb{N}}$ with $\vec{\xi}_i \in F$ for all $i \in \mathbb{N}$.

As in the case of the discrete spin system with finite range interaction we find again a subspace ℓ_r in the space $\ell(\Omega_>)$ which is mapped into itself by the above operator $\mathcal{L}$. Also the eigenfunction u to the highest eigenvalue λ_1 from Theorem II.2 will belong to this space ℓ_r . For this purpose we assume the functions M_j in the interaction Φ to be infinitely often differentiable on the space $F \times F$ which we assume to be a smooth manifold. Let $\ell(F^r, \mathbb{R})$ be the Banach space of all continuous real valued functions on the space $F^r := F \times .. \times F$. Then we define the space ℓ_r as folows:

$f \in \ell_r \Longleftrightarrow$ there exists a $g \in \ell(F^r, \mathbb{R})$ with $f(\xi_>) = g(\vec{\xi}_1, \ldots, \vec{\xi}_r)$ for all $\xi_> = (\vec{\xi}_i)_{i \in \mathbb{N}}$.

For $f \in \ell_r$ respectively the corresponding $g \in \ell(F^r, \mathbb{R})$ we get

$$\mathcal{L}g(\vec{\xi}_1, \ldots, \vec{\xi}_r) = \int_F d\omega \ g(\vec{x}, \vec{\xi}_1, \ldots, \vec{\xi}_{r-1}) \ \exp-\beta\left[h(\vec{x}) + \sum_{i=1}^{r} M_i(\vec{x}, \vec{\xi}_i)\right]. \quad (II.16)$$

This however tells us that the operator $\mathcal{L}$ when restricted to this space $\ell(F^r, \mathbb{R})$ is an integral operator with the kernel $\mathcal{L}(\underline{\xi}, \underline{\eta})$, $\underline{\xi} = (\vec{\xi}_1, \ldots, \vec{\xi}_r)$, $\underline{\eta} = (\vec{\eta}_1, \ldots, \vec{\eta}_r)$,defined as

$$\mathcal{L}(\underline{\xi}, \underline{\eta}) = \delta(\vec{\eta}_2 - \vec{\xi}_1) .. \delta(\vec{\eta}_r - \vec{\xi}_{r-1}) \ \exp-\beta\left[h(\vec{\eta}_1) + \sum_{i=1}^{r} M_i(\vec{\eta}_1, \vec{\xi}_i)\right],$$

$$(II.17)$$

where δ denotes the ordinary delta function.

We see therefore that the kernel $\mathcal{L}(.,.)$ for $r \geq 2$ is no longer a continuous function on the space $F^r \times F^r$ but in fact a highly singular distribution. So we can not apply directly the known theorems which would tell us immediately that the above integral operator is of trace class. So we have to look for another argument. For this we calculate

the kernel of the iterated integral operator $\mathcal{L}^r$: $\mathcal{C}(F^r, \mathbb{R}) \rightarrow \mathcal{C}(F^r, \mathbb{R})$.

After a straightforward and easy calculation where one uses only the

known properties of the δ-functions we get the regular expression

$$\mathcal{L}^r(\underline{\xi}, \underline{\eta}) = \exp{-\beta}\left[\sum_{i=1}^{r} h(\vec{\eta}_i) + \sum_{i=1}^{r-1} \sum_{j=1}^{r-i} M_j(\vec{\eta}_i, \vec{\eta}_{i+j}) + \right.$$

$$\left. + \sum_{i=1}^{r} \sum_{J=1}^{i} M_{r+j-i}(\vec{\eta}_i, \vec{\xi}_j)\right]. \qquad (II.18)$$

Since by assumption the functions M_i are $\mathcal{C}^\infty$ in both variables the

above kernel defines a nuclear operator of order zero in the Banach

space $\mathcal{C}(F^r, \mathbb{R})$. Comparing the above kernel of the operator $\mathcal{L}^r$ for a

system with interaction as given in (I.43) with expression (I.48) for

the Kramers-Wannier matrix for the same system we see that the two

are identical. The operator $\mathcal{L}$ in (I.47) is just the extension of the

operator $\mathcal{L}^r$ defined in the space $\mathcal{C}(F^r, \mathbb{R})$ to the larger Banach space

$\mathcal{L}_2(F^r, \mathbb{R})$. This extension is in fact uniquely determined because the

space $\mathcal{C}(F^r, \mathbb{R})$ is dense in the Hilbert space $\mathcal{L}_2(F^r, \mathbb{R})$.

Without going into the details we mention only that the operator $\mathcal{L}$

as defined in (II.16) is itself already a nuclear operator of order

zero when one restricts its domain of definition further in the way

we will discuss for other systems in chapter III.

Summarizing, we showed that also in the case of a continuous spin

system with a finite range interaction the Ruelle-Araki operator can

be regarded as the r-th root of the classical Kramers-Wannier transfer

matrix for such a system. This we achieved again by restricting the

original domain of definition for this operator to the subspace $\mathcal{C}_r$ of

the space $\mathcal{C}(\Omega_>)$.

Before coming in the next chapter to the more interesting systems

with long range interactions we want to apply the Ruelle-Araki trans-

fer matrix method to the continuous hard rod system with finite range

interaction. We will show that one gets this way for this system a
transfer matrix in the grand canonical ensemble whereas to our knowledge such a transfer matrix was known for this system only in the so
called pressure ensemble constructed by van Hove.

II.3. The continuous hard rod system with finite range interaction

The hard rod system is a system where extended particles of length
a move under the influence of an interaction Φ on the real axis $\mathbb{R}$.
The mathematical description of this system analogous to the one for
spin systems on a lattice given in the first chapter goes back to the
work of Gallavotti and Miracle-Sole [52] from where we took the following
notations. A configuration X of the infinite system is described by
a sequence of points $x_i \in \mathbb{R}$ which denote the coordinates of the position of a particle or more exactly of one endpoint of a particle.
Hence X can be identified with a subset of $\mathbb{R}$ which can be finite or
infinite depending on how many particles are present on the real line.

We denote then by Ω the space of all configurations of the infinite
system. A subset $Y \subset \mathbb{R}$ defines only then an allowed configuration
of the hard rod system if for all $y_i \in Y$ the inequality $|y_j - y_i| \geq a$
is fulfilled for $i \neq j$. This reflects just the fact that any two particles can not approach each other more than the distance a.

The empty set ϕ in $\mathbb{R}$ corresponds of course to the configuration
where no particle is present on the real line.

The space Ω can be made again a compact space by introducing the
following topology [52] :

Let X_n be a sequence of allowed configurations. Let $X_o \in \Omega$ be given.
Then we say X_n converges to X_o for $n \rightarrow \infty$ if for every finite interval
$[c,d]$, $c,d \notin X_o$, the configurations $X_n \cap [c,d]$ converge pointwise to
the configuration $X_o \cap [c,d]$.

The spaces $\Omega_>$ respectively $\Omega_<$ are again defined as the spaces of allowed configurations on the positive respectively negative real half

axis $\mathbb{R}_> = \{x \in \mathbb{R} : x \geq 0\}$ and $\mathbb{R}_< = \{x \in \mathbb{R} : x \leq 0\}$.

Similarly Ω_Λ denotes the space of all configurations on the fin-

ite interval Λ in $\mathbb{R}$.

The interactions Φ are defined as in the case of the lattice sys-

tems as continuous real valued functions on the space $\bigcup_{\Lambda \subseteq \mathbb{R}} \Omega_\Lambda$ with Λ

any finite subset of $\mathbb{R}$. We will consider also in this case only one-

and two-body interactions. Thereby we interpret as usual the one-body

interaction $\Phi(X) = \Phi_1(x)$ for $X = \{x\}$ as the chemical potential of the

particle at site x.

The grand partition function is then defined for a finite interval

$\Lambda = [0,L]$ in $\mathbb{R}$ as

$$Z_\Lambda = \int_{\Omega_\Lambda} dX \quad \exp{-\beta \, U_\Lambda^\Phi(X)} , \qquad (II.19)$$

where the symbol $\int_{\Omega_\Lambda} dX$ means simply

$$\int_{\Omega_\Lambda} dX = \sum_{n \geq 0} \int \dots \int 1/n! \, dx_1 \dots dx_n . \qquad (II.20)$$

The range of integration of every variable x_i in the above integral

is the interval Λ.

The energy function $U_\Lambda^\Phi(X)$ of the configuration $X = \{x_1, \dots, x_n\}$

is quite generally defined as

$$U_\Lambda^\Phi(X) = \sum_{j=1}^{n} \Phi_1(x_j) + \sum_{j=1}^{n-1} \sum_{k=j+1}^{n} \Phi_2(x_j, x_k) . \qquad (II.21)$$

The partition function with periodic boundary conditions reads then

$$Z_\Lambda = \int_{\Omega_\Lambda} dX \quad \exp{-\beta \left[U_\Lambda^\Phi(X) + W_{\Lambda, \mathbb{R} \setminus \Lambda}(X \cup Y) \right]} , \qquad (II.22)$$

where $W_{\Lambda, \mathbb{R} \setminus \Lambda}(X \cup Y)$ denotes the interaction energy of the configuration X in $\wedge$ with the configuration Y in $\mathbb{R}\backslash\wedge$ where $Y = \left\{ y \in \mathbb{R} : y = mL + x, m \in \mathbb{N}, x \in X \right\}$.

To be more precise one should call both the partition functions in (II.19) and (II.22) the configurational partition functions since we have completely omittet the contribution coming from the movement of the particles, that is the kinetic energy , to the Hamilton function. But this gives as is well known only some constant term which is not of importance for our discussion here. For convenience we will main- tain the notation used above for these partition functions.

The pressure $P_{\wedge}(\Phi)$ of the hard rod system with interaction Φ is defined according to the principles of statistical mechanics as

$$P_{\wedge}(\Phi) \;=\; 1/|\wedge| \quad \log \, z_{\wedge} \, . \qquad\qquad (II.23)$$

Theorem I.2 about the existence of the thermodynamic limit of lattice systems has a natural generalization for continuous systems. Without attaching great importance to give the most general conditions for the existence of this thermodynamic limit we recall here a result of Fisher [89] which reads in the case of the hard rod system as follows:

Theorem II.3 (Fisher) Assume the two-body interaction Φ_2 fulfills the conditions

a) $\Phi_2(x) \geqslant -b$ for all $x \in \mathbb{R}$ and

b) $|\Phi_2(x)| \leqslant c/|x|^{1+\varepsilon}, \varepsilon > 0$ as $|x| \to \infty$, c,b some constants,

then the mean pressure $p(\Phi)$ defined as

$$p(\Phi) : = \lim_{\wedge \to \infty} 1/|\wedge| \, \log \, z_{\wedge} = \lim_{\wedge \to \infty} P_{\wedge}(\Phi)$$

exists and is continuous in every finite dimensional space of parame- ters which describe the interaction Φ .

It was shown by Gallavotti and Miracle-Sole [52] that one can define also for such a hard rod system an operator $\mathcal{L}$ à la Ruelle-Araki in the space $\mathcal{C}(\Omega_{>})$ of observables of this system. The construction of this operator proceeds in analogy to the **lattice** case via conditional expectation values for the observable $f \in \mathcal{C}(\Omega_{>})$ so that we can omit the details here and give immediately its definition.

Namely, let $f \in \mathcal{C}(\Omega_{>})$ be an observable of the hard rod system . Then consider the linear operator $\mathcal{L}: \mathcal{C}(\Omega_{>}) \longrightarrow \mathcal{C}(\Omega_{>})$

$$\mathcal{L} f(X) : = \int_{Y \subset [0,a]} dY \; \exp{-\beta\left[U \overset{\Phi}{[0,a]}(Y) + W_{[0,a],\mathbb{R}\backslash[0,a]}(Y \cup X_a)\right]} f(Y \cup X_a),$$

$$(II.24)$$

where the symbol $\int_Y dY$ was already explained in (II.20). The configuration X_a in the above definition denotes the configuration one gets by shifting the configuration X the hard rod length a to the right, that means

$$X_a = \left\{ y \in \mathbb{R} : y = x + a, \; x \in X \right\} .$$

As mentioned already above the physical interpretation of this operator $\mathcal{L}$ is analogous to the one given for the Ruelle-Araki operator in the previous lattice case: $\mathcal{L} f$ is the conditional expectation value of the observable f with respect to the conditional Gibbs measure.

Also Theorem II.2 which established the connection between the operator $\mathcal{L}$ and the physical properties of a lattice system with interaction Φ has a natural extension to the case of a continuous hard rod system. For this purpose we assume the interaction Φ to be symmetric and translation invariant and fulfill the conditions

$\Phi_2(x) = \infty$ for $0 \leqslant x \leqslant a$ (hard rod potential),

$\Phi_2(.)$ is continuous for all $x \geqslant a$, $\qquad\qquad (II.25)$

$|\Phi_2(x)| \leqslant h(x)$, where h is some positive decreasing function on the real line with $\int_0^\infty (a + x) \, h(x) \, dx < \infty$.

Under these conditions on the interaction one can show [52]

Theorem II.4 (Gallavotti, Miracle-Sole)   Let the interaction Φ satisfy conditions (II.25). Let $\mathcal{L} : \ell(\Omega_>) \longrightarrow \ell(\Omega_>)$ be the linear operator defined in (II.24). Then there exist exactly one positive number λ_1, a positive function u in $\ell(\Omega_>)$ and a Borel measure γ on $\Omega_>$ with the following properties:

1) $\mathcal{L}u = \lambda_1 u$, $\mathcal{L}^*\gamma = \lambda_1 \gamma$, $\gamma(u) = 1$,where $\mathcal{L}^*$ denotes the dual operator of the operator $\mathcal{L}$ in the dual space $\ell(\Omega_>)^*$ of $\ell(\Omega_>)$.

2) For all $f \in \ell(\Omega_>)$, $f \neq 0$, the equality $\lim\limits_{n \to \infty} \| \lambda_1^{-n} \mathcal{L}^n f - \gamma(f)u \| = 0$ holds.

3) The mean pressure is given by $p(\Phi) = 1/\beta\, a \quad \log \lambda_1$.

4) The eigenvalue λ_1 and therefore also the pressure $p(\Phi)$ are continuously differentiable in Φ in every finite dimensional subspace of the parameter space.

5) The eigenfunction u can be written as

$$u(X) = c \int_{\Omega_<} \mu_<(dY) \quad \exp - \beta W_{\mathbb{R}_<, \, \mathbb{R}_>} (X \cup Y), \quad X \in \Omega_>,$$

where $\mu_<$ denotes again the Gibbs grand canonical measure on the negative real axis $\mathbb{R}_<$ and where c is some normalization constant.

We will next study this Ruelle-Araki transfer operator in more detail in the case of a finite range interaction Φ. Such an interaction Φ fulfills therefore the relation

$$\Phi_2(x) = 0 \quad \text{for all} \quad x \geq R\, a, \tag{II.26}$$

where R is some positive integer.

For this special case the operator $\mathcal{L}$ takes the following form

$$\mathcal{L} f(X) = f(X_a) + \int_0^{x_1 \wedge a} f(\{y\} \cup X_a) \ \exp{-\beta} \left[\Phi_1(y) + \sum_{\substack{x \in X \\ x \le Ra}} \Phi_2(x+a-y) \right] dy$$

$$(II.27)$$

where $x_1 = \min \{x \in X\}$ and $x_1 \wedge a = \min \{x_1, a\}$.

We look again for a subspace $\mathcal{L}_R$ of the space $\mathcal{L}(\Omega_>)$ of observables such that on the one hand the eigenvector u of Theorem II.4 belongs to $\mathcal{L}_R$ and on the other hand the operator $\mathcal{L}$ leaves this space invariant and has a simple spectrum when restricted to this space.

For this purpose we recall that for any configuration $X \in \Omega_>$, $X \ne \phi$, with $X = (x_1, x_2, ..)$ the points x_i must be such that $x_i \le x_{i+1} -a$ for all i. With this in mind we define the space $\mathbb{R}_+^R$ as

$$\mathbb{R}_+^R := \left\{ \vec{x} \in \mathbb{R}^R : \ 0 \le x_1 \le x_2 - a \le x_3 - 2a \le .. \le x_R - (R-1)a \right\}. \quad (II.28)$$

Denote by $\mathcal{L}_R(\mathbb{R}_+^R)$ the space of all continuous functions g on $\mathbb{R}_+^R$ with

$$g(x_1, .. x_R) = g(x_1, ..., x_{k-1}, Ra, (R+1)a, ..., (2R-k)a), \quad \text{if } k = \quad (II.29)$$
$$= \min \{i : x_i \ge R a\}.$$

This space $\mathcal{L}_R(\mathbb{R}_+^R)$ can be made a Banach space by introducing the following norm

$$\|g\| := \sup_{\vec{x} \in \mathbb{R}_+^R} |g(\vec{x})|. \quad (II.30)$$

We then define the subspace $\mathcal{L}_R$ of the space $\mathcal{L}(\Omega_>)$ as

$$\mathcal{L}_R := \left\{ f \in \mathcal{L}(\Omega_>) : \ \exists \ g \in \mathcal{L}_R(\mathbb{R}_+^R) : \ f(X) = g(x_1, ..., x_R) \ \text{if } |X| \ge R, \right.$$

$$f(X) = g(x_1, .., x_k, x_k + Ra, ..., x_k + (2R-k-1)a) \ \text{if } |X| = k < R \bigg\}. \quad (II.31)$$

Thereby the configuration X is given as $X = (x_1, x_2, \ldots)$.

It is obvious that the space $\mathcal{L}_R$ as defined above is a closed linear subspace of the Banach space $\mathcal{L}(\Omega_>)$. It remains to show that the operator $\mathcal{L}$ leaves this space invariant.

For this let $f \in \mathcal{L}_R$. In the case $|X| \geqslant R$ we get from (II.27)

$$\mathcal{L}f(X) = g(x_1 + a, \ldots, x_R + a) + \int_0^{x_1 \wedge a} g(y, x_1 + a, \ldots, x_{R-1} + a) \exp{-\beta \left[\bar{\Phi}_1(y) + \right.}$$

$$\left. + \sum_{i=1}^{R} \Phi_2(x_i + a - y) \right] dy \ . \qquad (II.32)$$

We have therefore to show that the function $h(x_1, \ldots, x_R)$ defined as

$$h(x_1, \ldots, x_R) := \int_0^{x_1 \wedge a} g(y, x_1 + a, \ldots, x_{R-1} + a) \exp{-\beta \left[\bar{\Phi}_1(y) + \right.}$$

$$\left. + \sum_{i=1}^{R} \Phi_2(x_i + a - y) \right] dy \ , \qquad (II.33)$$

belongs to the space $\mathcal{L}_R(\mathbb{R}_+^R)$, because then the function $f(X)$ can be written as

$$f(X) = g_a(x_1, \ldots, x_R) + h(x_1, \ldots, x_R) \qquad (II.34)$$

with both the functions $g_a(x_1, \ldots, x_R) := g(x_1 + a, \ldots, x_R + a)$ and $h(x_1, \ldots, x_R)$ in $\mathcal{L}_R(\mathbb{R}_+^R)$.

First of all it is clear that h is continuous in all the variables x_i. Assume furthermore $x_k \geqslant Ra$ for some k with $1 \leqslant k \leqslant R$ and $x_i < Ra$ for all $i < k$. Using the properties of the function $g \in \mathcal{L}_R(\mathbb{R}_+^R)$ we then get

$$h(x_1, \ldots, x_R) = \int_0^{x_1 \wedge a} g(y, x_1 + a, \ldots, x_{k-1} + a, Ra, \ldots, (2R-k-1)a) \exp{-\beta \left[\bar{\Phi}_1(y) + \right.}$$

$$\left. + \sum_{i=1}^{k-1} \Phi_2(x_i + a - y) + \sum_{i=k}^{R} \Phi_2(x_i + a - y) \right] dy \ .$$

Since $y \le a$ over the range of integration we have obviously for all

$i \ge k$: $x_i + a - y \ge R a$. This implies immediately

$$\sum_{i=k}^{R} \Phi_2(x_i+a-y) = \sum_{i=0}^{R-k} \Phi_2((R+i+1)a-y) \equiv 0$$

because $\Phi_2(x) \equiv 0$ for $x \ge R a$.

But this shows

$$h(x_1,..,x_R) = h(x_1,..,x_{k-1},Ra,(R+1)a,..,(2R-k)a) .$$

Let us next discuss the case $|X| = 1 < R$, $X \ne \Phi$.
This time we get

$$\mathcal{L} f(X) = g (x_1,..,x_1,x_1+Ra,..,x_1+(2R-1-1)a) + \int_{0}^{x_1 \wedge a} g(y,x_1+a,..,x_1+a,.)$$

$$\exp-\beta\left[\Phi_1(y) + \sum_{i=1}^{1}\Phi_2(x_i+a-y) + \sum_{i=0}^{R-1-1}\Phi_2(x_1+(R+i)a-y)\right] dy ,$$

$$(II.35)$$

where we used again the fact that $\sum_{i=0}^{R-1-1}\Phi_2(x_1+(R+i)a-y)$ vanishes
identically because of the finite range of the interaction Φ_2 .
The right hand side of (II.35) is however just the function (II.34)
taken at the point $(x_1,..,x_1,x_1+a,x_1+2a,..)$.

Summarizing the above we showed that the space ℓ_R is indeed mapped
by the operator $\mathcal{L}$ into itself. It is also easy to show that the
function u from Theorem II.4 belongs to the space ℓ_R.

This enables us now to discuss the operator $\mathcal{L}$ in the space $\ell_R(\mathbb{R}_+^R)$
and we can be sure of not having lost the eigenvalue λ_1 relevant for
the physical properties of the system expressed by the pressure $p(\Phi)$.

Let us mention another very important property of the operator $\mathcal{L}$
as defined in (II.24) which we will use immediately. For any $m \in \mathbb{N}$
the following relation is valid

$$\mathcal{L}^m = \mathcal{L}_{ma} ,$$

$$(II.36)$$

where the operator $\mathcal{L}_{ma}$ is defined as

$$\mathcal{L}_{ma} f(X) := \int\limits_{Y \subset [0,ma]} dY \quad f(Y \cup X_{ma}) \quad \exp{-\beta\left[U^{\Phi}_{[0,ma]}(Y) + \right.}$$

$$\left. + W_{[0,ma],\mathbb{R}\setminus[0,ma]}(Y \cup X_{ma}) \right] . \tag{II.37}$$

Thereby $U^{\Phi}_{[0,ma]}(Y)$ denotes the energy of the configuration Y on the interval $[0,ma]$ as introduced in (II.21) . The quantity $W_{[0,ma],\mathbb{R}\setminus[0,ma]}$ denotes the interaction energy of the configuration Y in the interval $[0,ma]$ and the configuration $X_{ma} := X + ma$ outside this interval. For $m = 1$ formula (II.36) is just the defining equation for the operator $\mathcal{L}$. The case $m > 1$ is proved by complete induction on m.

We are going to use formula (II.36) to determine the operator $\mathcal{L}^R$ in the space $\mathcal{L}_R(\mathbb{R}^R_+)$. This gives

$$\mathcal{L}^R g(x_1,\ldots,x_R) = g(x_1+Ra,\ldots,x_R+Ra) + \sum_{k=1}^{R} \int_0^{R_1(k)} dy_1 \ldots \int_{y_{k-1}}^{R_k(k)} dy_k .$$

$$\cdot \left\{ g(y_1,\ldots,y_k,x_1+Ra,\ldots,x_1+(R-k)a) \quad \exp{-\beta\left[\sum_{j=1}^{k} \Phi_1(y_j) + \right.} \right.$$

$$\left. + \sum_{j=1}^{k-1} \sum_{l=j+1}^{k} \Phi_2(y_1-y_j) + \sum_{j=1}^{R} \sum_{l=1}^{k} \Phi_2(x_j+Ra-y_1) \right] \right\}, \tag{II.38}$$

with $R_i(k) := \min\left\{ (R-(k-i+1))a+x_1; (R-k+i)a \right\}$, $1 \le i \le k$. Using then the properties of the functions $g \in \mathcal{L}_R(\mathbb{R}^R_+)$ as given in (II.29) we get

$$\mathcal{L}^R g(x_1,\ldots,x_R) = g(Ra,(R+1)a,\ldots,(2R-1)a) + \sum_{k=1}^{R} \int_0^{R_1(k)} dy_1 \ldots \int_{y_{k-1}+a}^{R_k(k)} dy_k .$$

$$\cdot \left\{ g(y_1,\ldots,y_k,Ra,(R+1)a,\ldots,(2R-k-1)a) \quad \exp{-\beta\left[\sum_{j=1}^{k} \Phi_1(y_j) + \right.} \right.$$

$$\left. + \sum_{j=1}^{k-1} \sum_{l=j+1}^{k} \Phi_2(y_1-y_j) + \sum_{j=1}^{R} \sum_{l=1}^{k} \Phi_2(x_j+Ra-y_1) \right] \right\}. \tag{II.39}$$

Without giving now the complete argument we mention that the operator $\mathcal{L}^{RN}$ can be shown to be a nuclear operator in the space $\mathcal{C}_R(\mathbb{R}^R_+)$ at least for $N \geq 2$. Unfortunately however its traces do not give the partition functions of our system. To get these partition functions we consider once more expression (II.27) for the operator $\mathcal{L}$. It is obvious that this can be written as

$$\mathcal{L} = \mathcal{L}_o + \mathcal{L}_1 \quad , \tag{II.40}$$

where $\mathcal{L}_o$ and $\mathcal{L}_1$ are defined as

$$\mathcal{L}_o f(X) : = f(X+a),$$

$$\tag{II.41}$$

$$\mathcal{L}_1 f(X) : = \mathcal{L} f(X) - \mathcal{L}_o f(X) .$$

We consider then the operator $\mathcal{L}_o \mathcal{L}^R$ in the space $\mathcal{C}_R(\mathbb{R}^R_+)$. For $g \in \mathcal{C}_R(\mathbb{R}^R_+)$ we get from expression (II.39) and using the definition (II.41) for the operator $\mathcal{L}_o$:

$$\mathcal{L}_o \mathcal{L}^R g(x_1,..,x_R) = g(Ra,..,(2R-1)a) + \sum_{k=1}^{R} \int_0^{\bar{R}_1(k)} dy_1 .. \int_{y_{k-1}+a}^{\bar{R}_k(k)} dy_k$$

$$\left\{ g(y_1,...,y_k,Ra,...,(2R-k-1)a) \quad \exp\text{-}\beta \sum_{j=1}^{k} \bar{\Phi}_1(y_j) + \right.$$

$$\left. + \sum_{j=1}^{k-1} \sum_{l=j+1}^{k} \bar{\Phi}_2(y_l-y_j) + \sum_{j=1}^{R} \sum_{l=1}^{k} \bar{\Phi}_2(x_j+(R+1)a-y_1) \right] \right\} , \tag{II.42}$$

with $\bar{R}_i(k) : = (R-k+i)a$ for all $1 \leq i \leq k$.

This shows that the operator $\mathcal{L}_o \mathcal{L}^R$ has a representation as

$$\mathcal{L}_o \mathcal{L}^R = \sum_{k=0}^{R} \tilde{\mathcal{L}}_k , \tag{II.43}$$

with $\tilde{\mathcal{L}}_o g(x_1,...,x_R) : = g(Ra,..,(2R-1)a)$

and

$$\tilde{\mathcal{L}}_k g(x_1,\ldots,x_R) = \int_0^{\overline{R}_1(k)} dy_1 \cdots \int_{y_{k-1}+a}^{\overline{R}_k(k)} dy_k \; g(y_1,\ldots,y_k,Ra,\ldots,(2R-k-1)a) \cdot$$

$$\cdot \left\{ \exp{-\beta\left[\sum_{j=1}^{k}\Phi_1(y_j) + \sum_{j=1}^{k-1}\sum_{l=j+1}^{k}\Phi_2(y_1-y_j) + \sum_{j=1}^{R}\sum_{l=1}^{k}\Phi_2(x_j+(R+1)a-y_1)\right]\right\}.$$

The operator $\mathcal{L}_o$ is trivially nuclear of order zero being a finite rank operator in the space $\mathcal{C}_R(\mathbb{R}_+^R)$. Its trace is given as

$$\text{trace}\,\mathcal{L}_o = 1 . \tag{II.44}$$

The operators $\tilde{\mathcal{L}}_k$, $1 \leq k \leq R$ on the other hand can themselves be written as

$$\tilde{\mathcal{L}}_k = T_k \circ S_k. \tag{II.45}$$

Thereby the operator $S_k : \mathcal{C}_R(\mathbb{R}_+^R) \longrightarrow \mathcal{C}_R(\mathbb{R}_+^k)$ is the restriction mapping

$$S_k g(x_1,\ldots,x_k) : = g(x_1,\ldots,x_k,Ra,\ldots,(2R-k-1)a) \tag{II.46}$$

and $T_k: \mathcal{C}_R(\mathbb{R}_+^k) \longrightarrow \mathcal{C}_R(\mathbb{R}_+^R)$ denotes the linear mapping

$$T_k g(x_1,\ldots,x_R) : = \int_0^{\overline{R}_1(k)} dy_1 \cdots \int_{y_{k-1}+a}^{\overline{R}_k(k)} dy_k \; g(y_1,\ldots,y_k) \cdot$$

$$\cdot \exp{-\beta\left[\sum_{j=1}^{k}\Phi_1(y_j) + \sum_{j=1}^{k-1}\sum_{l=1+1}^{k}\Phi_2(y_1-y_j) + \sum_{j=1}^{R}\sum_{l=1}^{k}\Phi_2(x_j+(R+1)a-y_1)\right]}.$$

$$\tag{II.47}$$

In case the functions Φ_1 and Φ_2 are smooth enough what we assume anyhow the operator T_k is a nuclear operator of order zero for every k as an operator from the space $\ell_R(\mathbb{R}_+^k)$ into the space $\ell_R(\mathbb{R}_+^R)$ [90].

Since the operators S_k are certainly bounded for every k this implies that the operator $\widetilde{\mathcal{L}}_k$ is in fact also nuclear of order zero [91]. Hence $\widetilde{\mathcal{L}}_k$ is of trace class and its trace is given as

$$\text{trace}\,\widetilde{\mathcal{L}}_k = \int_0^{(R-k+1)a} dy_1 \int_{y_1+a}^{(R-k+2)a} dy_2 \cdots \int_{y_{k-1}+a}^{Ra} dy_k \cdot$$

$$\cdot \exp{-\beta}\left[\sum_{j=1}^{k}\Phi_1(y_j) + \sum_{j=1}^{k-1}\sum_{l=j+1}^{k}\Phi_2(y_1-y_j) + \right.$$

$$\left. + \sum_{j=1}^{k}\sum_{l=1}^{k}\Phi_2(y_j+(R+1)a-y_1) + \sum_{j=0}^{R-k-1}\sum_{l=1}^{k}\Phi_2((R+j)a+(R+1)a-y_1)\right] \cdot$$

Since the argument of Φ_2 in the last sum of the above expression is always greater or equal to Ra this term does not contribute to the trace of $\widetilde{\mathcal{L}}_k$ and we get finally

$$\text{trace}\,\widetilde{\mathcal{L}}_k = \int_0^{(R-k+1)a} dy_1 \cdots \int_{y_{k-1}+a}^{Ra} dy_k \, \exp{-\beta}\left[\sum_{j=1}^{k}\Phi_1'(y_j) + \right.$$

$$\left. + \sum_{j=1}^{k-1}\sum_{l=j+1}^{k}\Phi_2(y_1-y_j) + \sum_{j=1}^{k}\sum_{l=1}^{k}\Phi_2(y_j+(R+1)a-y_1)\right] . \qquad (II.48)$$

It is now straightforward to convince oneself that expression (II.48) is just the contribution of all configurations to the partition function $Z_{(R+1)a}$ where there are exactly k particles present on the interval $[0, (R+1)a]$ with periodic boundary conditions, that means periodic repetition of the same configuration outside this interval. Summing these contributions over k and combining this with the trace of the operator $\widetilde{\mathcal{L}}_0$ which gives just the contribution of the empty set configuration we get therefore

<u>Lemma II.1</u> The trace of the operator $\mathcal{L}_0\mathcal{L}^R$ defined in (II.42) is identical to the grand partition function $Z_{(R+1)a}$ of our hard rod

system on the interval $\left[0, (R+1)a\right]$ with periodic boundary conditions.

After a quite long but in principle simple calculation in analogy to [92] one can then prove also

<u>Lemma II.2</u> The trace of the operator $\mathcal{L}_o \mathcal{L}^N$ is identical to the grand canonical partition function $Z_{(N+1)a}$ of the hard rod system on the interval $\left[0, (N+1)a\right]$ with periodic boundary conditions.

To summarize our discussion we therefore have shown that the hard rod system with a finite range interaction has a transfer matrix also in the grand canonical ensemble. The pressure $p(\beta, \Phi)$ as defined in Theorem II.3 can therefore be expressed by the highest eigenvalue of an operator as we will show immediately.

Indeed, according to Lemma II.2 we have

$$p(\beta, \Phi) = \lim_{n \to \infty} 1/na \quad \log Z_{na} = \lim_{n \to \infty} 1/na \ \log \text{trace} \ \mathcal{L}_o \mathcal{L}^{n-1}. \quad (II.49)$$

We see that we can not argue as simply as in the preceeding cases where the partition function Z_n was given as the trace of an operator $\mathcal{L}^n$. In the present case the argument goes as follows:

Define first a new operator $\mathcal{L}(z)$ as

$$\mathcal{L}(z) := z \mathcal{L}_o + \mathcal{L} \ , \quad\quad\quad\quad (II.50)$$

with $z \in \mathbb{C}$ and $\mathcal{L}_o$ respectively $\mathcal{L}$ as defined in (II.40) and (II.32). The operator $\mathcal{L}(z)$ defines a holomorphic operator [93] in the complexified space $\mathcal{L}_R(\mathbb{R}_+^R)$ and has furthermore the property that $\mathcal{L}(z)^{NR}$ is nuclear of order at least 2/3 for N large enough. The last property follows in fact from our previous discussions which can be generalized immediately to the operator (z).

For such a holomorphic family of trace class operators the following

formula is known to be valid [94]

$$(N+1) \ \text{trace} \ (\mathcal{L}_o \mathcal{L}(z)^N) = \frac{d}{dz} \ \text{trace} \ \mathcal{L}(z)^{N+1} \ . \tag{II.51}$$

At the point $z = 0$ we therefore get

$$\frac{d}{dz} \ \text{trace} \ \mathcal{L}(z)^{N+1} \Big|_{z=0} = (N+1) \ \text{trace} \ \mathcal{L}_o \mathcal{L}^N. \tag{II.52}$$

Applying on the other hand to the left hand side Grothendieck's

theorem [95], which says that the trace of a nuclear operator of order at

least 2/3 is given by the sum of its eigenvalues, we get

$$(N+1) \ \text{trace} \ \mathcal{L}_o \mathcal{L}^N = \frac{d}{dz} \sum_i \lambda_i^{N+1}(z) \Big|_{z=0} \ , \tag{II.53}$$

where $\{\lambda_i(z)\}$ are the non-vanishing eigenvalues of the operator $\mathcal{L}(z)$.

For z small enough we can apply to the operator $\mathcal{L}(z)$ Krasnoselskii's

theory of u_o-positive operators in a Banach space, which we recall in

Appendix C. This shows us that for real z the operator $\mathcal{L}(z)$ has a

simple and positive eigenvalue which is strictly greater than all

other eigenvalues of $\mathcal{L}(z)$ in absolute value. Denote this eigenvalue

by λ_1. Then λ_1 is real analytic for real z and even holomorphic for

a small neighbourhood of the real axis. Therefore we have [96]

$$\text{trace} \ \mathcal{L}_o \mathcal{L}^N = \lambda_1(0) \ \lambda_1'(0) + \frac{d}{dz} \sum_{i \neq 1} \lambda_i(z)^{N+1} \Big|_{z=0} \ , \tag{II.54}$$

where the second term on the right hand side of this expression cannot

be simplified in general. We assume however that the functions Φ_1

and Φ_2 in the interaction Φ are so smooth that the summation and the

differentiation can be interchanged in (II.54).

Doing so we get by inserting the result obtained into relation

(II.49)

$$p(\beta,\Phi) = \lim_{n\to\infty} 1/na \, \log(\lambda_1^{n-1} \, \lambda_1' + \sum_{i\neq 1} \lambda_i^{n-1} \, \lambda_i') =$$

$$= 1/a \, \log \lambda_1 \, . \tag{II.55}$$

But this shows that also in the case of the continuous hard rod system with finite range interaction the Ruelle-Araki operator defines a very useful transfer operator. This we achieved again by restricting its domain of definition from the space $\ell(\Omega_>)$ of observables of this system to a smaller space of observables which can be measured in finite intervals of the real line which are determined by the range of the interaction. In this space the Ruelle-Araki operator becomes then a nuclear operator with nice spectral properties.

We can therefore summarize our discussion up to this point as follows. We have seen that the Ruelle-Araki operator which was originally introduced by Ruelle and Araki as some abstract linear operator in the space of observables of such one-dimensional systems in fact can be regarded as a natural generalization of the classical concept of a trensfer matrix for these systems, as long as the interaction is of finite range. This identification can be achieved by simply restricting the domain of definition of this linear operator to certain subspaces of the original space $\ell(\Omega_>)$ of all observables of our system. In physical terms this restriction corresponds to the fact that because of the finite range of the interaction only observables which can be measured in finite regions are really of interest for such systems.

In the case of the continuous hard rod system the Ruelle-Araki operator defines a transfer matrix for this system also in the grand canonical ensemble which was not known before.

It is therefore not very surprizing that this operator is of some interest also in the case of long range interactions and has propert- ies which exceed also in this case by far those mentioned in the two Theorems II.2 and II.4 . The class of interactions where the program developped in this chapter can be fully carried out are the exponentia- lly decreasing interactions which we will discuss in the next chapter.

Thereby we will see that these interactions can be treated by the Ruelle-Araki transfer operator just like the finite range interactions.

III. SYSTEMS WITH LONG RANGE INTERACTIONS

We restrict our discussion to spin systems on a lattice with translation invariant interactions. For the sake of simplicity these interactions will be of one-body and two-body nature only. A generalization
to arbitrary n-body potentials of the method developped here should be
possible without much difficulties. A discussion of a continuous system with exponentially decreasing interactions ,namely the hard rod
system on the real line, was given in [92] .

To fix the notation we write the two-body part of the interaction
as

$$\phi_2(\xi_i, \xi_j) = -J(|j - i|) \; r(\xi_i, \xi_j) \; ,$$

where $r(.,.)$ is an arbitrary real valued function on the space $F \times F$.
Here F denotes as usual the set of possible spin values and $\xi = (\xi_i)$
is some configuration with $\xi_i \in F$ for all i .

III.1. Exponentially decreasing interactions with $J(i) = \exp{-\gamma i}, \gamma > 0$

Spin systems with this kind of interactions have been considered
the first time by M.Kac [97] and independently by G.Baker [98] . Their
studies have been followed by a series of papers [99] - [101] by themselves and other authors.

In the case of an Ising system that means for $F = \{1/2, -1/2\}$ and
$r(\xi_i, \xi_j) = \xi_i \cdot \xi_j$ Kac found an interesting ralation between this model with an exponentially decreasing interaction and the familiar Orn-
stein- Uhlenbeck process of probability theory. This relation allowed
him to reduce the calculation of the partition function for this system to spectral properties of a certain integral operator. Unfortunately however,the nature of this operator is only poorly understood
from a physical point of view. This comes from the fact that the re-

lation between the two systems mentioned above is more or less in the
dark [102]. So it is not clear if the operator considered by Kac is of
the same importance from the physics point of view as the classical
transfer matrix we discussed before for finite range interactions.

We will show that the discussion of the Ruelle-Araki transfer operator above for finite range interactions can be extended to include
also these long range interactions. So this operator will provide us
with a natural transfer matrix also for this case and resolves the
problems left by the method of Kac for these systems. The procedure
will be completely equivalent to the one we applied already for short
range interactions in the previous chapters.

Let us write then the interaction for the system (Z, F) in the
following form:

$$
\Phi(\xi_\Lambda) = \begin{cases} - h(\xi_i) & \text{if } \Lambda = \{i\} \subset z,\ \xi_\Lambda = (\xi_i), \\ - \lambda^{|j-i|}\ r(\xi_i, \xi_j) & \text{if } \Lambda = \{i,j\} \subset z,\ \xi_\Lambda = (\xi_i, \xi_j), \\ 0 & \text{otherwise.} \end{cases} \quad (III.0)
$$

Thereby we have written $\lambda = \exp - \gamma$ for $\gamma > 0$.

Let $\Omega_>$ and $\mathcal{C}(\Omega_>)$ be the configuration space respectively the
space of observables of our spin system. According to the general definition in (II.5) the action of the Ruelle-Araki operator $\mathcal{L}$ in the
space $\mathcal{C}(\Omega_>)$ of observables is given in the case of an interaction Φ
as in (III.0) by

$$
\mathcal{L} f(\xi_>) = \sum_{\sigma \in F} f(\sigma, \xi_>)\ \exp\beta\left[h(\sigma) + \sum_{i=1}^{\infty} r(\sigma, \xi_i)\ \lambda^i \right]. \quad (III.1)
$$

Thereby the configuration $\xi_>$ is given by $\xi_> = (\xi_i)_{i \in \mathbb{N}}$.

From Theorem II.2 we know that this operator $\mathcal{L}$ has a largest positive eigenvalue λ_1 with $\mathcal{L} u = \lambda_1 u$ which determines the free energy $f(\beta)$ of our system. Namely one has the relation $- \beta f(\beta) = \log \lambda_1$.

The eigenfunction u can be written as

$$u(\xi_{\rangle}) = c \int_{\Omega_{\langle}} \mu_{\langle}(d\xi_{\langle}) \; \exp\beta\left[\sum_{i=1}^{\infty} \sum_{j=1}^{\infty} r(\xi_{-i}, \xi_{j}) \; \lambda^{i+j}\right]. \qquad (III.2)$$

Our aim is to study this operator still more in detail as we did also in the case of finite range interactions. Our strategy will be exactly the same as in this last case. We will try to find a subspace $\mathcal{L}_{\infty}$ in the space $\mathcal{L}(\Omega_{\rangle})$ of observables of the system which should have again the following properties: first, the eigenvector u which deter- mines via its eigenvalue the physics should belong to this subspace. Second, the operator $\mathcal{L}$ should leave invariant this space and should have a nice simple spectrum when restricted to this space. We will see that this program can indeed be carried out.

To construct the space $\mathcal{L}_{\infty}$ we proceed as follows: if $d = |F|$ let $D_R := \left\{ z \in \mathbb{C}^d : |z_i| < R \text{ for } 1 \leq i \leq d \right\}$ be the open polycylinder in the complex space $\mathbb{C}^d$. We denote then by $A_{\infty}(D_R)$ the Banach space of all functions f holomorphic in D_R and continuous up to the closure $\bar{D}_R$ of D_R. The norm on this space is given by

$$\|f\| = \sup_{z \in D_R} |f(z)| < \infty \quad \text{for } f \in A_{\infty}(D_R).$$

Next define for $\xi_{\rangle} \in \Omega_{\rangle}, \xi_{\rangle} = (\xi_i)_{i \in \mathbb{N}}$ and $\sigma \in F$ a mapping $w_{\sigma} : \Omega_{\rangle} \to \mathbb{C}$ as

$$w_{\sigma}(\xi_{\rangle}) := \sum_{i=1}^{\infty} r(\sigma, \xi_i) \; \lambda^i . \qquad (III.3)$$

Let $\sigma_o := \max_{\sigma_i, \sigma_j \in F} |r(\sigma_i, \sigma_j)| .$

Then we have obviously for all $\sigma \in F$ and all configurations $\xi_{\rangle} \in \Omega_{\rangle}$ the estimate

$$|w_{\sigma}(\xi_{\rangle})| \leq \sigma_o \lambda/(1 - \lambda). \qquad (III.4)$$

Thereby we used the fact that $0 < \lambda < 1$.

Chose some $R_0 > \sigma_0 \lambda / (1 - \lambda)$. Then we define the following sub-space $\mathcal{L}_\infty$ of the space $\mathcal{L}(\Omega_>)$:

$$\mathcal{L}_\infty = \left\{ f \in \mathcal{L}(\Omega_>) : \exists\, g \in A_\infty(\,D_R) : \quad f(\xi_>) = g(\,w_{\sigma_1}(\xi_>),\ldots,w_{\sigma_d}(\xi_>)\,) \right\}. \quad (III.5)$$

One checks easily that the function u defined in (III.2) belongs to this space $\mathcal{L}_\infty$. To see this one defines a function $g = g(z_1,\ldots,z_d)$ as

$$g(z_1,\ldots,z_d) := \int_{\Omega_<} \mu(d\xi_<)\ \exp\beta \sum_{i=1}^{\infty} z_{\xi_{-i}} \lambda^i \ ,$$

with $z_{\sigma_i} := z_i$ for $1 \le i \le d$ and $F = (\sigma_1,\ldots,\sigma_d)$.

This function g is an entire function in $\mathbb{C}^d$ and belongs therefore trivially to the space $A_\infty(D_R)$.

Next we have to show that the operator $\mathcal{L}$ defined in (III.1) leaves the space $\mathcal{L}_\infty$ invariant.

Let $f \in \mathcal{L}_\infty$. Then we get

$$\mathcal{L} f(\xi_>) = \sum_{\sigma \in F} g(w_{\sigma_1}(\sigma,\xi_>),\ldots,w_{\sigma_d}(\sigma,\xi_>))\ \exp\beta\Big[h(\sigma) +$$
$$+ \sum_{i=1}^{\infty} r(\sigma,\xi_i)\,\lambda^i\Big]. \quad (III.6)$$

According to definition (III.3) for the mappings w_σ we get

$$w_{\sigma_k}(\sigma,\xi_>) = \lambda\, r(\sigma_k,\sigma) + \lambda w_{\sigma_k}(\xi_>) . \quad (III.7)$$

Let $\mathcal{W}_\sigma : D_{R_0} \longrightarrow D_{R_0}$ be a mapping defined as

$$(\mathcal{W}_\sigma)_i(z) := \lambda\, r(\sigma_i,\sigma) + \lambda z_i, \quad \sigma,\sigma_i \in F,\ 1 \le i \le d. \quad (III.8)$$

Because $R_0 > \sigma_0 \lambda / (1 - \lambda)$ it follows immediately that

$$\overline{\gamma_\sigma(D_{R_o})} \subset D_{R_o} \; , \tag{III.9}$$

that means γ_σ maps D_{R_o} strictly inside the region D_{R_o}. Furthermore, it is trivial that the mapping γ_σ is for every $\sigma \in F$ a holomorphic mapping.

Let $g \in A_\infty(D_{R_o})$ and define a function $\mathcal{L}g(z)$ as

$$\mathcal{L}g(z) := \sum_{k=1}^{d} g(\gamma_{\sigma_k}(z)) \; \exp\beta \left[h(\sigma_k) + z_k \right] . \tag{III.10}$$

The above arguments show that the function $\mathcal{L}g$ as defined in (III.10) is holomorphic in some open neighbourhood of the closed polycylinder $\overline{D_{R_o}}$ and belongs therefore especially to the Banach space $A_\infty(D_{R_o})$. Coming back to the function $\mathcal{L}f(\xi)$ in (III.6) we see that it can be written as

$$\mathcal{L}f(\xi) = \mathcal{L}g(w_{\sigma_1}(\xi),\ldots,w_{\sigma_d}(\xi)) . \tag{III.11}$$

But this shows that the operator $\mathcal{L}$ leaves the space ℓ_∞ invariant.

It makes therefore sense to restrict the domain of definition of the operator $\mathcal{L}$ to this space and consider it from now on as a linear operator in the Banach space ℓ_∞ or what amounts to the same in the Banach space $A_\infty(D_{R_o})$. Its definition there is given in (III.10).

As shown in [106] one can apply Grothendiecks theory of nuclear operators in Banach spaces to the type of operators as given by our operator $\mathcal{L}$. In fact, one has the following result, which we formulate as a theorem:

Theorem III.1 (Ruelle) Let D be an open bounded domain in the complex space $\mathbb{C}^d$. Let $\psi: D \to D$ be a holomorphic mapping such that $\overline{\psi(D)} \subset D$. Denote by $A_\infty(D)$ the Banach space of functions holomorphic in D with the sup norm. For $\varphi \in A_\infty(D)$ define the linear operator $T : A_\infty(D) \longrightarrow A_\infty(D)$ as

$$Tf (z) : = \varphi(z) \quad f \circ \psi(z).$$

Then one has

1) the mapping ψ has exactly one fixed point z^* in D,

2) the operator T is nuclear of order zero,

3) the trace of T is given by the formula

$$\text{trace } T = \varphi(z^*) \quad \det(1 - \psi'(z^*))^{-1},$$

where $\psi'(z^*)$ is the Fréchet derivative of the mapping ψ at the point $z = z^*$.

We will not give the proof of this theorem here because it turned out that it is only a special case of a more general theorem which we are going to discuss in Appendix B. We refer therefore to this appendix. For a discussion of Grothendieck's theory of nuclear operators we refer to Appendix A where we collected the main points of this theory.

We can now apply Theorem III.1 to our operator $\mathcal{L}$ as given in (III.10). This operator is therefore of trace class in the space $A_\infty(D_{R_o})$. This means we have in fact achieved our first goal namely getting an operator with simple nice spectral properties.

By applying the trace formula of the above Theorem III.1 we get after a trivial computation

$$\text{trace } \mathcal{L} = 1/(1 - \lambda)^d \sum_{\sigma \in F} \exp \beta \left[h(\sigma) + \lambda/(1-\lambda) \ r(\sigma,\sigma) \right]. \quad \text{(III.12)}$$

But this is up to the factor $1/(1 - \lambda)^d$ just the partition function Z_1 of our system for one lattice site and periodic boundary conditions. This is so simple that we need not give any more details about the calculation.

Let us see how this can be generalized to the system with arbitrary
lattice sites. We want to show

Corollary III.1   The operator $\mathscr{L}^n : A_\infty(D_{R_o}) \to A_\infty(D_{R_o})$ is nuclear
of order zero and its trace is given by the formula

$$\text{trace } \mathscr{L}^n = (1 - \lambda^n)^{-d} \; z_n \;,$$

where z_n is the partition function for n sites of the spin system
with exponentially decresing interaction ϕ as in (III.0) with periodic
boundary conditions.

Proof: That the operator $\mathscr{L}^n$ is again nuclear of order zero follows
from the general theory of nuclear operators as discussed in Appendix
A. To show the trace formula we proceed as follows: we write the
operator $\mathscr{L}$ as a finite sum $\mathscr{L} = \sum_{\sigma \in F} \mathscr{L}_\sigma$ where the meaning of the
operators $\mathscr{L}_\sigma$ is clear from the definition (III.10) of $\mathscr{L}$. This induces for the operator $\mathscr{L}^n$ the decomposition

$$\mathscr{L}^n = \sum_{i_1=1}^{d} \cdots \sum_{i_n=1}^{d} \mathscr{L}_{\sigma_{i_1}} \circ \cdots \circ \mathscr{L}_{\sigma_{i_n}} \;.$$

We use the abbreviation

$$T_{(j_1 \ldots j_n)} := \mathscr{L}_{\sigma_{j_n}} \circ \cdots \circ \mathscr{L}_{\sigma_{j_1}} \;.$$

We will calculate then the trace of this operator.
By complete induction on n we get for the action of the operator
$T_{(j_1 \ldots j_n)}$ on some element $g \in A_\infty(D_{R_o})$ the expression

$$T_{(j_1 \ldots j_n)} \, g(z) = g(\{\sum_{k=1}^{n} \lambda^k r(\sigma_{j_k}, \sigma_i) + \lambda^n \, z_i\}) \; \exp\beta\Big[\sum_{k=1}^{n} h(\sigma_{j_k}) +$$

$$+ \sum_{s=1}^{n-1} \sum_{k=1}^{n-s} \lambda^s \, r(\sigma_{j_k}, \sigma_{j_{k+s}}) + \sum_{k=1}^{n} \lambda^n \, z_{j_k} \Big] \;. \qquad \text{(III.13)}$$

The partition function Z_n on the other hand for the interval $[1,n]$ in $Z_>$ for our system with periodic boundary conditions can be written as

$$Z_n = \sum_{j_1=1}^{d} \cdots \sum_{j_n=1}^{d} Z_{(j_1..j_n)} , \qquad \text{(III.14)}$$

where the quantity $Z_{(j_1..j_n)}$ denotes the contribution to the partition function Z_n stemming from the configuration $\xi_>(\sigma_{j_1}, \ldots, \sigma_{j_n})$. This configuration is defined as follows

$$\xi_>(\sigma_{j_1}, \ldots, \sigma_{j_n}) = (\xi_i)_{i \in \mathbb{N}} \quad \text{with} \quad \xi_i = \sigma_{j_i} \text{ for } 1 \leq i \leq n \quad \text{and}$$

$$\xi_{i+n} = \xi_i \text{ for all } i \in \mathbb{N} .$$

According to our discussion in (I.28) we get for $Z_{(j_1..j_n)}$

$$Z_{(j_1 \ldots j_n)} = \exp\beta\Big[\sum_{k=1}^{n} h(\sigma_{j_k}) + \sum_{k=1}^{n-1} \sum_{l=k+1}^{n} \lambda^{l-k}/(1-\lambda^n)\ r(\sigma_{j_k}, \sigma_{j_l}) +$$

$$+ \sum_{k=1}^{n} \lambda^n/(1-\lambda^n)\ r(\sigma_{j_k}, \sigma_{j_k}) +$$

$$+ \sum_{k=2}^{n} \sum_{l=1}^{k-1} \lambda^{n+1-k}/(1-\lambda^n)\ r(\sigma_{j_k}, \sigma_{j_l}) \Big]. \qquad \text{(III.15)}$$

We will compare this expression with the one we get for the trace of the operator $T_{(j_1..j_n)}$ from relation (III.13). This is done again with the trace formula of Theorem III.1 :

$$\text{trace } T_{(j_1..j_n)} = 1/(1-\lambda^n)^d \left\{ \exp\beta\Big[\sum_{k=1}^{n} h(\sigma_{j_k}) + \sum_{s=1}^{n-1} \sum_{k=1}^{n-s} \lambda^s . \right.$$

$$\left. \cdot r(\sigma_{j_k}, \sigma_{j_{k+s}}) + \sum_{k=1}^{n} \sum_{l=1}^{n} \lambda^{n+1-k}/(1-\lambda^n)\ r(\sigma_{j_k}, \sigma_{j_l})\Big] \right\} =$$

$$1/(1-\lambda^n)^d \left\{ \exp\beta \left[\sum_{k=1}^{n} h(\sigma_{j_k}) + (1-\lambda^n)^{-1} \sum_{s=1}^{n-1} \sum_{k=1}^{n-s} \lambda^s \cdot \right. \right.$$

$$\left. \cdot r(\sigma_{j_k}, \sigma_{j_{k+s}}) - (1-\lambda^n)^{-1} \sum_{s=1}^{n-1} \sum_{k=1}^{n-s} \lambda^{n+s} \; r(\sigma_{j_k}, \sigma_{j_{k+s}}) + \right. \qquad \text{(III.16)}$$

$$\left. \left. + (1-\lambda^n)^{-1} \sum_{k=1}^{n} \sum_{l=1}^{n} \lambda^{n+1-k} \; r(\sigma_{j_k}, \sigma_{j_l}) \right] \right\}.$$

The last term in this formula we will rewrite still in a different way as

$$\sum_{k=1}^{n} \sum_{l=1}^{n} \lambda^{n+1-k} \; r(\sigma_{j_k}, \sigma_{j_l}) = \sum_{k=1}^{n} \lambda^n \; r(\sigma_{j_k}, \sigma_{j_k}) +$$

$$\qquad \text{(III.17)}$$

$$+ \sum_{k=2}^{n} \sum_{l=1}^{k-1} \lambda^{n+1-k} \; r(\sigma_{j_k}, \sigma_{j_l}) + \sum_{k=1}^{n-1} \sum_{l=k+1}^{n} \lambda^{n+1-k} \; r(\sigma_{j_k}, \sigma_{j_l}).$$

Because furthermore

$$\sum_{s=1}^{n-1} \sum_{k=1}^{n-s} \lambda^{s+n} \; r(\sigma_{j_k}, \sigma_{j_{k+s}}) = \sum_{k=1}^{n-1} \sum_{l=k+1}^{n} \lambda^{n+1-k} \; r(\sigma_{j_k}, \sigma_{j_l}) \quad ,$$

we get after inserting expression (III.17) into (III.16) :

$$\text{trace } T_{(j_1..j_n)} = 1/(1-\lambda^n)^d \left\{ \exp\beta \left[\sum_{k=1}^{n} h(\sigma_{j_k}) + 1/(1-\lambda^n) \cdot \right. \right.$$

$$\left. \cdot \sum_{s=1}^{n-1} \sum_{k=1}^{n-s} \lambda^s \; r(\sigma_{j_k}, \sigma_{j_{k+s}}) + 1/(1-\lambda^n) \sum_{k=1}^{n} \lambda^n \; r(\sigma_{j_k}, \sigma_{j_k}) + \right.$$

$$\left. \left. + 1/(1-\lambda^n) \sum_{k=2}^{n} \sum_{l=1}^{k-1} \lambda^{n+1-k} \; r(\sigma_{j_k}, \sigma_{j_l}) \right] \right\}.$$

But this is up to the factor $(1-\lambda^n)^{-d}$ just the expression (III.15) we found for the quantity $Z_{(j_1..j_n)}$. Summation over the multiindex $(j_1,...,j_n)$ then proves the assertion of Corollary III.1 .

Thus by restricting the domain of definition of the Ruelle-Araki operator to the space $\mathcal{L}_\infty$ we achieved exactly what we wanted: in this space the operator $\mathcal{L}$ is the transfer matrix of our system in the sense of Kramers and Wannier. We get therefore for the free energy $f(\beta)$

$$-\beta f(\beta) = \lim_{n\to\infty} 1/n \ \log \ (1 - \lambda^n)^d \sum_{\{i\}} \lambda_i^n \ , \qquad (III.18)$$

where λ_i are the eigenvalues of the operator $\mathcal{L}$ in the space $A_\infty(D_{R_0})$ counted according to their algebraic multiplicities. The sum in the above expression (III.18) exists because the operator $\mathcal{L}$ and all its iterates $\mathcal{L}^n$ are nuclear of order zero (see Theorem A.2 of Appendix A).

Because the eigenvector u of Theorem II.2 belongs to the space $\mathcal{L}_\infty$ respectively to the space $A_\infty(D_{R_0})$ it follows from this theorem that the highest eigenvalue of the operator $\mathcal{L}$ in the space $A_\infty(D_{R_0})$ is just the one mentioned in this last theorem. So far our method reproduces only this result of Ruelle summarized in Theorem II.2 .

We want to show however more. Theorem II.2 makes no statement about the rest of the spectrum of the operator $\mathcal{L}$ which determines on the other hand certainly the analyticity properties of the different eigenvalues and so by relation (III.18) also those of the free energy of our spin system. It turns out that the highest eigenvalue λ_1 of the Ruelle-Araki transfer matrix $\mathcal{L}$ when restricted to the space $A_\infty(D_{R_0})$ is a simple eigenvalue and completely separated in absolute value from the rest of the spectrum of $\mathcal{L}$ which consists indeed of eigenvalues only. This means that for any $\varsigma \in \sigma(\mathcal{L})$, $\varsigma \neq \lambda_1$, one has $\lambda_1 > \varsigma$.

Ruelle succeeded to show such a property for exponentially decreasing interactions by a detailed study of the operator $\mathcal{L}$ in the space $\ell(\Omega_>)$ [107] . One drawback of his method is however that he does not get any statement concerning the rest of the spectrum of $\mathcal{L}$ besides the highest eigenvalue λ_1. One knows on the other hand that a knowledge of the complete spectrum of $\mathcal{L}$ would be necessary to describe for instance also the correlation functions of the spin system.

The method we are going to develop here allows a much deeper understanding of the spectrum of the transfer matrix $\mathcal{L}$, certainly only on
the restricted domain of definition $A_\infty(D_{R_o})$. But it is just on this
space where the Ruelle-Araki operator becomes the true transfer matrix
in the sense of Kramers and Wannier. The spectrum of this operator in
the space $\ell(\Omega_>)$ considered by Ruelle is not of great physical importance besides its highest eigenvalue.

Another advantage of our method will be that we can give simple
formulas for calculating the two highest eigenvalues of the operator
$\mathcal{L}$ in the Banach space $A_\infty(D_{R_o})$ which remind us of the Ritz variational
principle in Hilbert spaces. Such prescriptions do not exist in Rue-
lle's analysis of the problem.

We saw in the case of systems with finite range interactions that
the Ruelle-Araki transfer matrix became a trace class operator when
restricted in its domain of definition to certain subspaces of the space of all observables $\ell(\Omega_>)$. Its spectrum was determined first of
all by its positivity properties. The theorems of Perron-Frobenius
respectively of Jentzsch described then the characteristic properties
of the highest eigenvalue of the transfer operator. It is therefore
not so astonishing that similar things happen also in the present case
of exponentially decreasing interactions. It is known that the theory
of such positive operators can be generalized to arbitrary Banach spaces so that the above mentioned cases appear only as very special examples of a much more deeper theory.

The main features of this theory which was developped especially by
russian mathematicians like Krein and Rutman [108] respectively in the
form we are using it here by Krasnoselskii and Ladyzenskii [109], [110]
we have described in Appendix C to which we refer the reader for a
better understanding of the discussion which follows.

III.1.1. Spectral properties of the operator $\mathcal{L}$

The most interesting properties of our operator $\mathcal{L}$ when restricted to the space $A_\infty(D_{R_0})$ follow from Theorem C.2 of Appendix C. To apply this theorem we have to verify the assumtions made in the theorem. To do so we have to identify first of all corresponding quantities in the theorem and in our operator $\mathcal{L}$. We set

$$D := D_{R_0} \quad \text{and} \quad \psi_i(z) := \gamma_{\sigma_i}(z) \,. \qquad (III.20)$$

Furthermore denote by $\varphi_i(z)$ the function

$$\varphi_i(z) := \exp\beta\left[h(\sigma_i) + z_i\right], \qquad (III.21)$$

where the quantities D_{R_0} and γ_{σ_i} had been defined in (III.5) respectively (III.8).

We have to show that the mappings γ_{σ_i} are separating in the sense defined in Appendix C. Let $z^*_{\sigma_i}$ be the fixed points of these mappings. From their definition in (III.8) we then get

$$(z^*_{\sigma_i})_k = \lambda/(1-\lambda) \ r(\sigma_k, \sigma_i) \,, \quad k,i = 1,..,d \,.$$

Therefore all these points belong to the set $B_{R_0} := D_{R_0} \cap \mathbb{R}^d \subset \mathbb{R}^d$ independent of i. We then consider only such functions $r(.,.): F \times F \to \mathbb{R}$ which have the property that the set $\left\{\psi^{\alpha_1,\cdots,\alpha_d}(z)\right\}$ is a set of uniqueness for every $z \in B_{R_0}$ and for every $f \in A_\infty(D_{R_0})$ [111]. But this just means that the mappings γ_{σ_i} are indeed separating. For a definition of the mappings $\psi^{\alpha_1\cdots\alpha_d}_i$ see Appendix C. An example for an allowed function $r(.,.)$ is any function symmetric in its arguments and for which $r(\sigma_i, \sigma_j) \neq r(\sigma_k, \sigma_l)$ for $i \neq k$ and $j \neq l$.

The inequality for the highest eigenvalue λ_1, given in Theorem C.2

in Appendix C reads in the case of the transfer operator $\mathcal{L}$:

$$\max_{z \in \bar{B}_{R_0}} \left(\sum_{\sigma \in F} \exp \beta \left[h(\sigma) + z_\sigma \right] \right) \geq \lambda_1 \geq \min_{z \in \bar{B}_{R_0}} \left(\sum_{\sigma \in F} \exp \beta \left[h(\sigma) + z_\sigma \right] \right) . \quad (III.22)$$

Furthermore it follows from the same theorem that this eigenvalue is in absolute value larger than any of the other eigenvalues of the operator $\mathcal{L}$. Standard perturbation theory then shows that this eigenvalue is analytic in all parameters which enter the operator itself in an analytic manner. An example for such a parameter is certainly the number β which is just proportional to inverse temperature.

Let us next try to understand the operator $\mathcal{L}$ and its spectrum still better. Especially we want to derive a formula which allows us later to determine the highest eigenvalues in an explicit way.

For reasons of economy in writing not too much we will restrict our discussion from now on to the simplest case of a spin 1/2 Ising system with exponentially decreasing interaction without an external field. Completely analogous considerations can certainly be carried out also in the general scheme we discussed up to this point .

The interaction we consider from now on is therefore given as

$$\phi(\xi_\Lambda) = \begin{cases} - J \lambda^{|j-i|} \xi_i \xi_j & \text{for } \xi_\Lambda = (\xi_i, \xi_j), \Lambda = \{i,j\} \subset \mathbb{Z}, \\ \\ 0 & \text{otherwise.} \end{cases} \quad (III.23_1)$$

One can then regard the corresponding transfer matrix $\mathcal{L}$ as a linear operator $\mathcal{L}: A_\infty(D_R) \longrightarrow A_\infty(D_R)$, where D_R is now the open disc of radius R in the complex plane $\mathbb{C}$. The radius R has to be chosen in the appropriate way as we will see immediately. If we then take the set F as $F = \{1, -1\}$ the Ruelle-Araki transfer operator $\mathcal{L}$ looks like

$$\mathcal{L}g(z) = \exp \beta Jz \; g(\lambda + \lambda z) + \exp{-\beta Jz} \; g(-\lambda + \lambda z) . \quad (III.23)$$

We see therefore that this operator maps the space $A_\infty(D_R)$ really inside itself if the radius R of the disc D_R is chosen such that $R > \lambda/(1-\lambda)$ [105].

One verifies also without difficulties that the above operator $\mathcal{L}$ leaves the following two subspaces A_∞^+ and A_∞^- of the space $A_\infty(D_R)$ invariant:

$$A_\infty^+ := \left\{ g \in A_\infty(D_R) : g(z) = g(-z) \right\}$$

$$(III.24)$$

$$A_\infty^- := \left\{ g \in A_\infty(D_R) : g(z) = -g(-z) \right\}.$$

From the definition (III.23) of the operator $\mathcal{L}$ one derives that the function $g(z) := f(-z)$ is an eigenfunction of the operator $\mathcal{L}$ with eigenvalue ρ if the function $f(z)$ itself is such an eigenfunction to the same eigenvalue. Therefore the linear independent eigenfunctions to any eigenvalue can be taken always from the spaces A_∞^+ and A_∞^- : One only has to take the symmetric respectively the antisymmetric linear combination of the functions $f(z)$ and $f(-z)$.

According to Theorem C.2 the eigenfunction f_1 which belongs to the highest eigenvalue λ_1 is positive on the set $\overline{B}_R$. Because furthermore this highest eigenvalue is simple we conclude that the function f_1 must belong to the space A_∞^+ . Therefore this function must be a symmetric function in z . Let us discuss some further properties of this function f_1.

III.1.1.1. Properties of eigenvectors of the operator $\mathcal{L}$

Let n be any natural number. We then denote by $A_n^+(D_R)$ the Banach space of all symmetric functions $f(z)$ holomorphic in D_R with

$$\| f \|_n := \sup_{|\alpha| \leq n} \sup_{z \in D_R} | D^\alpha f(z) | < \infty \quad . \qquad (III.25)$$

Thereby we denoted by $\underline{\alpha}$ the multiindex $\underline{\alpha} = (\alpha_1, \alpha_2)$, $\alpha_i \in \mathbb{N} \cup \{0\}$ and by $D^{\underline{\alpha}}$ the differential operator

$$D^{\underline{\alpha}} = \frac{\partial^{|\underline{\alpha}|}}{\partial x^{\alpha_1} \partial y^{\alpha_2}} \quad ,$$

where we have written $z = x + i y$ and $|\underline{\alpha}| = \alpha_1 + \alpha_2$.

The functions $f \in A_n^+(D_R)$ have continuous derivatives on the closed disc $\overline{D}_R$ up to order n.

Theorem III.1 can be carried over without difficulties to the space $A_n^+(D_R)$ and shows that the operator $\mathcal{L}$ defined in (III.23) when re-stricted to the space $A_n^+(D_R)$ is again a nuclear operator of order zero and has in this space similar properties as in the space A_∞^+. This one can see already from the fact that for any $g \in A_\infty^+$ the function $\mathcal{L}g$ belongs already to the space $A_n^+(D_R)$ for any n . The function $\mathcal{L}g$ namely is holomorphic in some open neighbourhood of the closure of the disc D_R and therefore continuously differentiable in this neighbourhood.

From this it follows particularly that all symmetric eigenfunctions of the operator $\mathcal{L}$ must belong to any of the spaces $A_n^+(D_R)$.

An analogous statement is certainly true for the antisymmetric eigenfunctions. They all belong to any of the spaces $A_n^-(D_R)$ which are defined similar to the spaces $A_n^+(D_R)$ only that the functions must be antisymmetric in the argument z.

Define next a real Banach space $A_{2n}^{+,\mathbb{R}}(D_R)$ which will be of some importance for our following discussion of the positivity properties of the operator

$$A_{2n}^{+,\mathbb{R}}(D_R) \; : = \; \left\{ f \in A_{2n}^+(D_R) \; : \; f \text{ is real on the interval } [-R,R] \right\}.$$

In this space we define the following cone K_n:

$$K_n : = \left\{ f \in A_{2n}^{+,\mathbb{R}}(D_R) : f^{(2j)}(0) \geq 0 \quad \text{for all } 0 \leq j \leq n \text{ and} \right.$$

$$\left. f^{(2n)}(x) \geq 0 \quad \text{for all } x \in [-R,R] \right\}. \qquad \text{(III.26)}$$

It is clear that the cone K_n is proper and reproducing in the real Banach space $A_{2n}^{+,\mathbb{R}}(D_R)$. This means just that any $f \in A_{2n}^{+,\mathbb{R}}(D_R)$ can be written as $f = g_1 - g_2$ with $g_i \in K_n$ for $i = 1, 2$. To show this property we define the function $g(z) : = \cosh z$. Then $g(z)$ belongs certainly to the cone K_n for all $n \in \mathbb{N}$.

Let $M_1 : = \max\limits_{0 \leq j \leq n} f^{(2j)}(0)$ and $M_2 : = \max\limits_{-R \leq x \leq R} f^{(2n)}(x)$.

We denote then by $M : = \max (M_1, M_2)$. But then it is trivial that the function $f(z) + M g(z)$ belongs again to the cone K_n. Denote this element of the cone by $\tilde{g}$. But this shows that f can be written as $f(z) = \tilde{g}(z) - M g(z)$ and this is just what we wanted to show.

To apply the theory of positive operators as outlined in Appendix C we have to show next that the operator $\mathcal{L}$ in (III.23) is indeed positive, that means leaves the cone K_n invariant.

Let $f \in K_n$. Then we get for $0 \leq j \leq n$ and all $x \in [-R,R]$

$$(\mathcal{L}f)^{(2j)}(x) = \sum_{k=0}^{j} \binom{2j}{2k} \lambda^{2j-2k} (\beta J)^{2k} (\exp \beta Jx \quad f^{(2j-2k)}(\lambda + \lambda x) +$$

$$+ \exp - \beta Jx \quad f^{(2j-2k)}(-\lambda + \lambda x)) +$$

$$\qquad \qquad \qquad \qquad \text{(III.27)}$$

$$+ \sum_{k=0}^{j-1} \binom{2j}{2k+1} \lambda^{2j-2k-1} (\beta J)^{2k+1} (\exp \beta Jx \quad f^{(2j-2k-1)}(\lambda + \lambda x) -$$

$$- \exp - \beta Jx \quad f^{(2j-2k-1)}(-\lambda + \lambda x)).$$

Because $f^{(s)}(x) \geq 0$ for all $f \in K_n$, all $0 \leq s \leq 2n$ and all $x \in [0,R]$ it follows from relation (III.27) that

$$(\mathcal{L} f)^{(2j)} (0) \geq 0 \quad \text{for all } j . \tag{III.28}$$

From the monotonicity property of the functions $f^{(j)}(x)$, $0 \leq j \leq 2n-1$, for positive x it follows that

$$\left| f^{(j)}(x) \right| \leq \left| f^{(j)}(y) \right|$$

for all $x,y \in [-R,R]$ with $y \geq x$. Taking into account the fact that the functions $f^{(2k)}(x)$ are symmetric for all $0 \leq k \leq n$ we deduce that $f^{(2k)}(x) \geq 0$ for all $x \in [-R,R]$ and all $0 \leq k \leq n$. This implies immediately that the first sum in expression (III.27) is non-negative for all $x \in [-R,R]$. For positive x it follows from the same arguments that also the first term in the second sum is non-negative. Now one has for positive x the trivial inequality $\lambda + \lambda x \geq \left| -\lambda + \lambda x \right|$. This and the above mentioned monotonicity of $f^{(k)}(x)$ for $0 \leq k \leq 2n-1$ shows that

$$\exp \beta Jx \ f^{(2j-2k-1)}(\lambda + \lambda x) \geq \left| \exp -\beta Jx \ f^{(2j-2k-1)}(-\lambda + \lambda x) \right| .$$

This on the other hand implies immediately for all $0 \leq j \leq n$

$$(\mathcal{L} f)^{(2j)} (x) \geq 0 \quad \text{for all } x \in [-R,R] . \tag{III.29}$$

But this shows that the cone K_n is indeed left invariant by the operator $\mathcal{L}$ and therefore this operator is a positive operator in the real Banach space $A_{2n}^{+,\mathbb{R}}(D_R)$.

To apply finally Theorem C.1 of Krasnoselskii as given in Appendix C we still have to show that the operator is even u_0-positive with respect to the cone K_n (see Appendix C for the relevant definitions).

This we will do in two steps. First we show that the operator $\mathcal{L}$ is u_0-upper bounded: this means that there exists an $u_0 \in K_n$, $u_0 \neq 0$, such that for all $f \in K_n$ with $f \neq 0$ there exist a number $p \in \mathbb{N}$ and a real number $\alpha > 0$ with

$$\mathcal{L}^p f \leq \alpha u_0 . \tag{III.30}$$

Let $u_0 = \cosh z$. Then $u_0 \in K_n$ and $u_0 \neq 0$. For any $f \in K_n$ with $f \neq 0$ let M be a positive number such that $M \geq \max_{0 \leq j \leq n-1} (\mathcal{L}f)^{(2j)}(0)$ and at the same time $M \geq \max_{x \in [-R,R]} (\mathcal{L}f)^{(2n)}(x)$.

Trivially the inequality

$$(\mathcal{L} f) \leq M u_0 \tag{III.31}$$

is then valid. But this means just that $\mathcal{L}$ is u_0-upper bounded with respect to the cone K_n.

In a second step we show that the operator $\mathcal{L}$ is also u_0-lower bounded with respect to the cone K_n. We have therefore to show that there exists an element $u_0 \in K_n$, $u_0 \neq 0$ such that we can find for every $f \in K_n$, $f \neq 0$, numbers $q \in \mathbb{N}$ and $\beta > 0$ such that

$$\mathcal{L}^q f \geq \beta u_0 .$$

We use once more formula (III.27). We claim the number $(\mathcal{L}f)^{(2j)}(0)$ cannot vanish for any j with $0 \leq j \leq n$. Assume there exists such an j with $(\mathcal{L}f)^{(2j)}(0) = 0$. Because all the terms in expression (III.27) are seperately positive we must have in particular $f(\lambda) = 0$. But f is a monotone increasing function for $x \geq 0$ which is non-negative by definition. Therefore this function must identically vanish for all x from the interval $0 \leq x = \lambda$. On the other hand f was a holomorphic

function in the disc D_R and must therefore vanish identically in the

domain of holomorphy. This is in contradiction to our assumption $f \neq 0$.

Therefore our claim, namely $f^{(2j)}(0) \neq 0$ for all $0 \leq j \leq n$, is

correct. Using the monotonicity argument again then shows that for

all $x \in [-R,R]$ and all $0 \leq j \leq n-1$ we have $(\mathcal{L}f)^{(2j)}(x) \neq 0$.

It remains to show that also $(\mathcal{L}f)^{(2n)}(x) > 0$ on the whole interval

$-R \leq x \leq R$. Assume again there exists a number y with $0 \leq y \leq R$

such that $(\mathcal{L}f)^{(2n)}(y) = 0$. The same reasoning as above shows again

that in formula (III.27) all terms must vanish for $x = y$. In parti-

cular this leads to $f(\lambda + \lambda y) = 0$. As above the monotonicity argu-

ment applies and one deduces again $f \equiv 0$.

Let therefore m be the minimum of the two numbers m_1 and m_2 which

are given as

$$m_1 := \min_{0 \leq j \leq n-1} (\mathcal{L}f)^{(2j)}(0)$$

and

$$m_2 := \min_{-R \leq x \leq R} (\mathcal{L}f)^{(2n)}(x) \; / \; \max_{-R \leq x \leq R} \cosh x \;.$$

According to the arguments given above we have $m > 0$ and also

$$\mathcal{L}f \geq m u_o, \tag{III.32}$$

where u_o denotes again the function $\cosh z$.

This proves that the operator $\mathcal{L}$ is also u_o-lower bounded with respect

to the cone K_n in the space $A_{2n}^{+,\mathbb{R}}(D_R)$.

Relations (III.31) and (III.32) together show now that the operator

$\mathcal{L}$ is in fact u_o-positive with respect to the cone K_n where u_o de-

notes the function $\cosh z$ $\boxed{110}$.

Now we can really apply Theorem C.1 of Appendix C. According to it the eigenvector f_1 belonging to the highest eigenvalue λ_1 belongs to the cone K_n. Therefore this eigenvector f_1 has the property that for all $k \in \mathbb{N} \cup \{0\}$ and all $x \in [-R,R]$ the relations

$$f_1^{(2k)} (x) > 0 \quad \text{and} \quad f_1 (x) = f_1 (-x) \tag{III.33}$$

are valid.

This follows immediately from relation (III.32) with $f = f_1$ and the fact that $\cosh x > 0$ on the whole real axis.

The same analysis given above for the eigenvector f_1 in the space $A_{2n}^{+,\mathbb{R}} (D_R)$ is true also in the space $A_{2n}^{+} (D_R)$ as shown in $[135]$.

Without giving all the details here we mention only that the eigenvector f_2 belonging to the second highest eigenvalue λ_2 can be characterized in exactly the same way: one considers the operator $\mathcal{L}$ now in the space $A_{2n+1}^{-} (D_R)$ respectively in the real Banach space $A_{2n+1}^{-,\mathbb{R}} (D_R)$. This space is defined in analogy to the space $A_{2n}^{+,\mathbb{R}} (D_R)$, the only difference is that the functions are now antisymmetric instead of being symmetric as in the former case. Next one considers again a certain cone K_n^- in the space $A_{2n+1}^{-,\mathbb{R}} (D_R)$:

$$K_n^- : = \left\{ f \in A_{2n+1}^{-,\mathbb{R}} (D_R): \quad f^{(2j+1)} (0) \geq 0 \quad \text{for all } 0 \leq j \leq n \quad \text{and} \right.$$

$$\left. f^{(2n+1)} (x) \geq 0 \quad \text{for all} \quad x \in [0,R] \right\}. \tag{III.34}$$

Taking for u_0 the function $\sinh z$ which certainly belongs to K_n^- it is straightforward to see by the same method as in the symmetric case that the operator $\mathcal{L}$ is u_0-positive also in the space $A_{2n+}^{-,\mathbb{R}}$ with respect to the above cone K_n^-. Krasnoselskii's Theorem shows therefore that also the highest eigenvalue $\tilde{\lambda}_2$ of the operator $\mathcal{L}$ in the space $A_{2n+1}^{-,R}(D_R)$ is simple, positive and strictly larger than all other eigenvalues of this operator in this space in absolute value. The eigenvector f_2 belonging to this highest eigenvalue $\widetilde{\lambda}_2$ belongs to the cone K_n^- and is therefore in particular an antisymmetric function in z. Because the eigenfunction f_1 corresponding to the eigenvalue λ_1 in the symmetric space does definitely not belong to K_n^- we see immediately that $\lambda_1 > \widetilde{\lambda}_2$. To see that $\widetilde{\lambda}_2 \geqslant |\varsigma|$ for all $\varsigma \in \sigma(\mathcal{L})$, $\varsigma \neq \lambda_1, \lambda_2$ where we denoted by $\sigma(\mathcal{L})$ the spectrum of the operator $\mathcal{L}$ in the space $A_\infty(D_R)$ which contains all the above spaces as subspaces, we look at the special case J = 0. In this case the operator $\mathcal{L}$ becomes trivial in the sense that its spectrum can be determined explicitly. Thereby it turns out that the eigenvector belonging to the second largest eigenvalue λ_2 is in fact antisymmetric in z. Because both the eigenvalies λ_1 and $\widetilde{\lambda}_2$ according to Krasnoselskii's Theorem are analytic in J the eigenvalue $\widetilde{\lambda}_2$ found in the space $A_{2n}^{-,R}(D_R)$ must be identical to the eigenvalue λ_2 of the operator $\mathcal{L}$ in the larger space $A_\infty(D_R)$. The argument for changing to the complex Banach space proceeds exactly like in [135].

In analogy to expression (III.33) we find therefore for the eigenfunction corresponding to the eigenvalue λ_2 which we denoted by f_2 the following properties:

$$f_2^{(2k+1)}(x) \geq 0 \quad \text{for } k \in \mathbb{N} \cup \{0\} \text{ and } f_2(x) = - f_2(-x) \qquad \text{(III.35)}$$

for all $x \in [-R,R]$.

So far our discussion of some properties of the eigenvectors of the transfer matrix $\mathcal{L}$ belonging to the two highest eigenvalues. Next we turn our attention to some properties of these eigenvalues themselves. In fact, they are of primary interest because of their connexion with the physical properties of our spin system as we have seen already.

Our aim thereby will be to find a reasonable simple formula for an explicit calculation of these eigenvalues. This will be a mini-max principle well known from the Hilbert space theory of linear operators.

III.1.1.2. Properties of the highest eigenvalues of the operator $\mathcal{L}$

In this section we come back to our starting point and consider the operator $\mathcal{L}$ again in the space $A_\infty(D_R)$.

From the proof of Theorem C.2 in Appendix C we know that the operator $\mathcal{L}$ as defined in (III.23) is u_o-positive with respect to the cone K_o in the real Banach space $A_\infty^{\mathbb{R}}(D_R) = \left\{ f \in A_\infty(D_R) : f \text{ is real on } [-R,R] \right\}$. For u_o one could take the function $u_o = 1$: then there exist to every $f \in K_o$, $f \not\equiv 0$, numbers $p \in \mathbb{N}$ and $\alpha, \beta > 0$ such that $\beta u_o \leq \mathcal{L}^p f \leq \alpha u_o$. The cone K_o was thereby defined as

$$K_o = \left\{ f \in A_\infty^{\mathbb{R}}(D_R) : \quad f(x) \geq 0 \quad \text{for } -R \leq x \leq R \right\} .$$

For every $g \in K_o$ with $g(x) > 0$ the following is obviously true

$$(\min_{-R \leq x \leq R} g(x)) \, u_o \leq g \leq (\max_{-R \leq x \leq R} g(x)) \, u_o \tag{III.36}$$

Therefore the operator $\mathcal{L}$ is also g-positive for any such g from the open kernel $\overset{o}{K}_o$ of the cone K_o. Let u then be any function from $\overset{o}{K}_o$. The function $\mathcal{L}u(x)/u(x)$ is obviously a well defined continuous function on the interval $-R \leq x \leq R$ and takes there both its maximum and minimum. But this way we get

$$\min_{-R \leq x \leq R} (\mathcal{L}u(x)/u(x)) \, u \leq \mathcal{L}u \leq \max_{-R \leq x \leq R} (\mathcal{L}u(x)/u(x)) \quad . \tag{III.37}$$

From this follows however by using Krasnoselkii's theorem

$$\min_{-R \leq x \leq R} \mathcal{L}u(x)/u(x) \leq \lambda_1 \leq \max_{-R \leq x \leq R} \mathcal{L}u(x)/u(x) \ . \qquad (III.38)$$

Because this is true for all $u \in \mathring{R}_o$ it follows that

$$\sup_{u \in \mathring{R}_o} \min_{-R \leq x \leq R} \mathcal{L}u(x)/u(x) \leq \lambda_1 \leq \inf_{u \in \mathring{R}_o} \max_{-R \leq x \leq R} \mathcal{L}u(x)/u(x) \ . \quad (III.39)$$

According to Theorem C.2 the eigenvector f_1 belonging to the highest eigenvalue λ_1 is an element of the cone K_o , it belongs even to $\mathring{R}_o$. Setting therefore $u = f_1$ in relation (III.39) we get

$$\max_{u \in \mathring{R}_o} \min_{-R \leq x \leq R} \mathcal{L}u(x)/u(x) = \lambda_1 = \min_{u \in \mathring{R}_o} \max_{-R \leq x \leq R} \mathcal{L}u(x)/u(x) \ .(III.40)$$

We see that the highest eigenvalue of the operator $\mathcal{L}$ in the space $A_\infty(D_R)$ can be determined via a principle which is well known from the Hilbert space theory of linear operators.

It is therefore also not surprising that an analogous principle allows also to calculate the second highest eigenvalue. It is not necessary to give all the details of the arguments which in fact are completely analogous to the case just discussed. Instead of the cone K_o one has to take the cone K_o^- in the real Banach space $A_1^{-,\mathbb{R}}(D_R)$ as defined in (III.34). We denote its open kernel again by $\mathring{R}_o^-$. Then one finds

$$\min_{u \in \mathring{R}_o^-} \max_{-R \leq x \leq R} (\mathcal{L}u)\dot{}(x)/u\dot{}(x) = \lambda_2 = \max_{u \in \mathring{R}_o^-} \min_{-R \leq x \leq R} (\mathcal{L}u)\dot{}(x)/u\dot{}(x) \ .$$

$$(III.41)$$

We are going to apply these formulas in the next section for calculating the highest eigenvalues in a simple spin system, namely the already mentioned Kac model.

III.1.2. The van der Waals limit of the Kac model

We use relation (III.40) for a simple proof for the existence of
a phase transition in an Ising system with exponentially decreasing
interaction in the so called van der Waals limit. This limit is defined in the following way [55]: we let the coupling constant J in the
interaction ϕ as given in (III.23$_1$) tend to zero in such a way that
the product of J with the range of interaction of ϕ stays constant.
This means that one takes the limit $\gamma \to 0$ of an interaction ϕ_γ of the
form

$$\phi_\gamma(\xi_\Lambda) = - J_0\gamma \ \xi_i \ \xi_j \ \exp{-\gamma |j-i|} \ , \qquad\qquad (III.42)$$

for $\xi_\Lambda = (\xi_i, \xi_j)$ and $\Lambda = \{i,j\} \subset \mathbb{Z}_> $.
It is thereby important to take the limit $\gamma \to 0$ after the thermodynamic
limit $\Lambda \to \infty$ because otherwise the system becomes a free system and
therefore trivial. This van der Waals limit for the above interaction
ϕ_γ was discussed besides others by Kac [99] and independently also
by Baker [98]. Later the discussion was taken up again now in a much
more general set-up by Lebowitz and Penrose [112]. They showed that
this van der Waals limit is a way to derive in a rigorous way the
van der Waals equation of state, respectively in the case of a spin
system the classical Curie-Weiss theory of magnetism.

In a series of papers several authors tried to understand the behavior of the above system (III.42) also in the so called critical region that means for γ infinitesimal small but different from zero.

But to our knowledge this has not been done up to now in a mathematical satisfactory manner [101]. The method applied in these investigations is just simple minded perturbation theory in the variable γ.

Unfortunately however nobody really has any idea how far such a perturbation expansion really makes sense. One sees namely from the definition of the operator $\mathcal{L}$ in (III.23) that the limit $\gamma \to 0$ leads to an
operator which is no longer of trace class and which therefore can have
a very complicated spectrum so that an expansion in γ seems very pro-
blematic. A confirmation for this our scepticism is for instance the
fact that a naive perturbation calculation in γ leads for all $\gamma \neq 0$
to a highest eigenvalue $\lambda_1(\beta)$ which is non analytic in β $\left[55\right]$.

For $\gamma \neq 0$ the eigenvalue $\lambda_1(\beta)$ must however be an analytic function in β as we have seen in the preceding sections.

We believe that perhaps our method for treating such systems with
an interaction decreasing exponentially fast at infinity can shed a
new light on these problems. But this problem of the behavior in the
critical region should not be persued further in this work here and we
leave it as an open problem for the future.

What we want to do here is to use our methods developped in the
last section to prove non-analyticity of $\lambda_1(\beta)$ in the limit $\gamma = 0$.
This can be achieved by a simple application of relation (III.40).

To show this it is useful to consider instead of the operator $\mathcal{L}$
in (III.23) the following slightly different operator $\widetilde{\mathcal{L}}_\lambda : A_\infty(D_1) \to A_\infty(D_1)$
with

$$\widetilde{\mathcal{L}}_\lambda f(z) = \exp \beta\left[\jmath z/(1-\lambda)\right] f(\lambda(1-\lambda) + \lambda z) +$$

$$\text{(III.43)}$$

$$+ \exp-\beta\left[\jmath z/(1-\lambda)\right] f(-\lambda(1-\lambda) + \lambda z)$$

and D_1 the unit disc in $\mathbb{C}$.

The operators $\mathcal{L}_\lambda$ are defined for all $0 \leq \lambda < 1$ on the same
Banach space $A_\infty(D_1)$ and have there furthermore the same properties
as shown before for the operators $\mathcal{L}_\lambda$ where we have added the dependence

on λ also for the operator $\mathcal{L}$. Furthermore it is also trivial that the operators $\mathcal{L}_\lambda$ and $\tilde{\mathcal{L}}_\lambda$ have exactly the same spectrum for $0 \leq \lambda < 1$. We can now introduce the interaction Φ_γ as in (III.42). For this we replace in expression (III.43) the constant J by the quantity $J_o (1 - \lambda)$ which for γ very small leads exactly to the transfer operator for a system with interaction Φ_γ. Proceeding this way we get the operator

$$\mathcal{L}_\lambda f(z) = \exp \beta J_o z \quad f(\lambda (1 - \lambda) + \lambda z) +$$

(III.44)

$$+ \exp - \beta J_o z \quad f(- \lambda (1 - \lambda) + \lambda z) .$$

For $\lambda \neq 1$ we can apply the results of the previous section and get according to relation (III.40) for the highest eigenvalue $\lambda_1 = \lambda_1 (\lambda)$:

$$\max_{u \in \mathring{R}_o} \min_{x \in I} \mathcal{L}_\lambda u(x) / u(x) = \lambda_1 (\lambda) = \min_{u \in \mathring{R}_o} \max_{x \in I} \tilde{\mathcal{L}}_\lambda u(x) / u(x), \quad (III.45)$$

where I denotes the interval $[0,1]$ in $\mathbb{R}$ and $\mathring{R}_o$ is defined in analogy to the previous section.

From (III.45) we get the following estimates for λ_1 :

$$\lambda_1 (\lambda) \geq \min_{x \in I} \tilde{\mathcal{L}}_\lambda u(x) / u(x)$$

respectively

(III.46)

$$\lambda_1 (\lambda) \leq \max_{x \in I} \tilde{\mathcal{L}}_\lambda u(x) / u(x),$$

where u is any function from the interior $\mathring{R}_o$ of the cone K_o. For this function we will take now the test functions $g_i (z)$, $i = 1, 2, 3$ given by

$$g_1 (z) := 1,$$

$$g_2(z) : = \cosh (J_o \beta z (1 - \lambda)^{-1}) , \tag{III.47}$$

$$g_3(z) : = \exp (\alpha z^2 / (1 - \lambda)) , \alpha \in \mathbb{R} . \tag{III.48'}$$

Using then function g_1 we get from (III.46)

$$2 \cosh \beta J_o \geq \lambda_1 \geq 2. \tag{III.48}$$

Function g_2 on the other hand gives after a simple calculation

$$\tilde{\mathcal{L}}_\lambda g_2(z) / g_2(z) = \exp \beta J_o \lambda + \exp - \beta J_o \lambda (\exp[\beta J_o(1-2\lambda)(1-\lambda)^{-1}x] +$$

$$+ \exp - [\beta J_o(1-2\lambda)(1-\lambda)^{-1}x]) (\exp[\beta J_o(1-\lambda)^{-1}x] +$$

$$+ \exp - [\beta J_o(1-\lambda)^{-1}x])^{-1} ,$$

and therefore

$$2 \cosh \beta \lambda J_o \geq \lambda_1 \geq \exp \beta J_o \lambda . \tag{III.49}$$

Consider finally the function g_3 which gives

$$\tilde{\mathcal{L}}_\lambda g_3(x) / g_3(x) = 2 \exp \alpha [\lambda^2 (1 - \lambda)] \exp - \alpha [x^2(1+\lambda)] \cosh ((\beta J_o + 2\lambda^2 \alpha) x) . \tag{III.50}$$

Denoting the right hand side of this last expression by $h(x)$ we conclude that the function $h(x)$ behaves like $\exp - \alpha x^2 (1+\lambda)$ for $|x| \to \infty$ and vanishes therefore very fast at infinity.

For the behavior of the function $h(x)$ for small x on the other hand one gets

$$h(x) \sim 2 \exp(\alpha \lambda^2 (1-\lambda)) \quad (1 + c(\alpha, \lambda) x^2), \tag{III.51}$$

with

$$c(\alpha, \lambda) = -\alpha (\lambda + 1) + (J_0 \beta)^2 / 2 + 2 \lambda^2 \alpha \beta J_0 + 2 \lambda^4 \alpha^2 .$$

For $\lambda = 1$ which is just the van der Waals limit this leads to

$$c(\alpha, 1) = 2((\alpha - (1 - J_0 \beta)/2)^2 + (2 J_0 \beta - 1)/4) . \tag{III.52}$$

For $\alpha = (1 - J_0 \beta)/2$ and $J_0 \beta < 1/2$ this gives therefore $c(\alpha, 1) < 0$. If one choses now $\lambda - 1$ sufficiently small a simple continuity argument shows that

$$c(\alpha, \lambda) < 0 \quad \text{for} \quad \alpha = (1 - J_0 \beta)/2 \quad \text{and} \quad J_0 \beta < 1/2. \tag{III.53}$$

This now leads for such values of α and $J_0 \beta$ for small enough $1 - \lambda$ to the following upper bound for the eigenvalue λ_1:

$$\lambda_1 \leq 2 \exp \alpha \lambda^2 (1 - \lambda) . \tag{III.54}$$

Comparing this result with relation (III.48) in the limit $\lambda = 1$ we deduce that for $J_0 \beta < 1/2$ the eigenvalue must be $\lambda_1 = 2$. On the other hand however it follows from estimate (III.49) that $\lambda_1 \geq \exp \beta J_0$. This shows already that the function $\lambda_1 (\beta)$ cannot be analytic in the variable β. But this means that there exists a phase transition in this limit $\lambda = 1$.

This result was found already by Kac and Baker by different methods. We wanted to show only that the method we are using here in this simple example is perhaps less complicated and nevertheless completely rigorous. Perhaps this simplicity will allow in the furure to apply it also to the unsolved problems in connection with this van der Waals limit in exponentially decreasing interactions. A possible starting point thereby could be the two formulas (III.40) and (III.41) .

Before we are coming in the next section to a discussion of continuous spin systems on a lattice with exponentially decreasing interactions we want to consider briefly some generalizations of the Ising model, namely the N-state Potts model. This model is for instance useful in describing the following situation [113] , [114] : Consider a system on the lattice consisting of N different kinds of atoms. Every lattice site can be occupied by exactly one atom . Atoms on different lattice sites can interact with each other only if they are of the same kind. This is just the situation arising in certain alloys. Now, this interaction can be described as follows if we denote the set of different atoms by $F = \{ \vartheta_i , \quad 1 \leq i \leq N \}$:

$$\phi (\xi_\wedge) = - J \, \lambda^{|i-j|} \, \delta (\xi_i , \xi_j) \tag{III.55}$$

for $\xi_\wedge = (\xi_i , \xi_j)$, $\wedge = \{ i, j \} \subset \mathbb{Z}$, $\xi_i \in F$.

In expression (III.55) we used the abbreviation $\delta (\xi_i , \xi_j)$ which is a generalized Kronecker symbol defined as

$$\delta (\vartheta_i , \vartheta_j) = \delta_{ij} \qquad \text{for} \quad 1 \leq i, j \leq N. \tag{III.56}$$

Comparing the form (III.55) of the interaction with the general form given in (III.0) for the interactions for which our whole analysis is valid, we see that the interaction of the N-state Potts model belongs to this class. We only have to identify the function $r(.,.)$ of expression (III.0) with the function given in (III.56).

We can therefore apply the whole apparatus of section III.1. also to the Potts model. The treatment is in fact simpler than the one we gave for this model in [115] and has the advantage of being the same for all such discrete spin models.

We will next discuss continuous spin systems on a lattice with exponentially decreasing interactions.

III.1.3. Continuous spin systems on a lattice

We consider the case where F is any compact manifold with a finite measure $d\omega$. As before we denote then the elements of F by $\vec{x}$. The interaction Φ should be again of the form given in (III.O), where now the $\vec{\xi}_i$'s are continuous variables in the space F. The Ruelle-Araki operator $\mathcal{L}$ in the space $\ell(\Omega_{>})$ has then the following form:

$$\mathcal{L}f(\xi_{>}) = \int_F d\omega \ f(\vec{x},\xi_{>}) \ \exp\beta\left[h(\vec{x}) + \sum_{i=1}^{\infty} r(\vec{x}, \vec{\xi}_i) \ \lambda^i\right] , \qquad (III.57)$$

when the configuration $\xi_{>}$ is again given as $\xi_{>} = (\vec{\xi}_i)_{i\in\mathbb{N}}$.

As in the foregoing cases one can try again to find a subspace ℓ_{∞} in the space $\ell(\Omega_{>})$ of observables which has similar nice properties as we found them in the case of discrete spin variables. However it turns out that this cannot be achieved so easily as before. Only for certain functions $r(.,.)$ in the interaction Φ the procedure can be carried through without difficulties. We want to scetch briefly the problems arising in the general case.

We start completely analogous to our discussion in the discrete case. Let $\ell(F)$ be the Banach space of all continuous functions in the space F with the sup-norm and denote by B_R the open ball of radius R in this space, that means

$$B_R := \left\{ z \in \ell(F) : \|z\| < R \right\}. \qquad (III.58)$$

We then define a continuous mapping $w : \Omega_{>} \to \ell(F)$ by

$$w(\xi_>)\ (\vec{x}) \ : \ = \ \sum_{i=1}^{\infty} \ r(\vec{x},\vec{\xi}_i)\ \lambda^i \ . \tag{III.59}$$

Let $\sigma_o \ : \ =_{\vec{x},\vec{y}\in F} \max \ |r(\vec{x},\vec{y})|.$ Then we have obviously

$$|w(\xi_>)\ (\vec{x})| \ \leq \ \sigma_o \ \lambda/(1-\lambda)$$

and therefore also

$$\|w(\xi_>)\| \leq \sigma_o \ \lambda/(1-\lambda) \qquad \text{for all} \ \ \xi_>\in\Omega_> \ .$$

Therefore the set $\{w(\xi_>) \ : \ \xi_>\in\Omega_> \}$ of functions is compact in the space $\mathcal{C}(F)$ and is contained strictly inside the ball B_{R_o} if R_o is chosen for instance in such a way that $R_o > \sigma_o \ \lambda/(1-\lambda) \ .$

Denote then by $A_\infty(B_{R_o})$ the Banach space of all holomorphic bounded functions in the ball B_{R_o}. (For properties of functions holomorphic over infinite dimensional domains see for instance $[115]$ or $[116]$.)

One could then try to construct the subspace $\mathcal{L}_\infty$ in analogy to our procedure in section III.1. as follows:

$$f\in\mathcal{L}_\infty \Leftrightarrow \exists g\in A_\infty(B_{R_o}) \ : \ f(\xi_>) = g(\ w(\xi_>)) \ . \tag{III.60}$$

It is easy to show that the operator $\mathcal{L}$ as defined in (III.57) leaves indeed this space invariant. Namely, take a f from $\mathcal{L}_\infty$ and let $g\in A_\infty(B_{R_o})$ be such that $f\ (\xi_>) = g(w(\xi_>))$. Define another function $\mathcal{L}g(z)$ as

$$\mathcal{L}g(z) \ : \ = \ \int_F d\omega \ g(\psi_{\vec{x}}(z)) \ \exp\beta\left[h(\vec{x}) + z(\vec{x})\right], \tag{III.61}$$

where $\psi_{\vec{x}} \ : \ B_{R_o} \longrightarrow B_{R_o}$ denotes the holomorphic mapping

$$\psi_{\vec{x}}(z)\ (\vec{y}) \ : \ = \ \lambda \ r(\vec{y},\vec{x}) + \lambda z(\vec{y}) \ , \qquad \vec{x},\vec{y}\in F \ .$$

The function $\mathcal{L}g$ so defined is obviously again an element of the space $A_\infty(B_{R_O})$. It is furthermore immediately to verify that

$$\mathcal{L}f(\xi_{\rangle}) = \mathcal{L}g(w(\xi_{\rangle})) . \qquad\qquad (III.62)$$

This allows us again to restrict the domain of definition of the operator $\mathcal{L}$ to this space $\mathcal{L}_\infty$ respectively to the space $A_\infty(B_{R_O})$ of holomorphic functions over B_{R_O}. This is a reasonable thing to do because also the eigenfunction belonging to the highest eigenvalue of the operator $\mathcal{L}$ in the space $\mathcal{C}(\Omega_{\rangle})$ belongs to this space as can be easily verified. So far nothing exciting happened in comparison with the discrete spin case .

Unfortunately it turns now out that the operator $\mathcal{L}$ even when resticted to the space $A_\infty(B_{R_O})$ as defined in (III.61) is not yet a trace class operator. This can be seen as follows. If we calculate formally the trace of $\mathcal{L}$ analogous to the formula of Theorem B.1 of Appendix B we get (formally!)

$$\text{trace}\,\mathcal{L}= \int_F d\omega \ \exp\beta\left[h(\vec{x}) + z_{\vec{x}}(\vec{x})\right] \ \det\ (1-\psi_{\vec{x}}^{'}\ (z_{\vec{x}}^{*}))^{-1}, \qquad (III.63)$$

where $z_{\vec{x}}^{*}$ denotes the fixed point of the mapping $\psi_{\vec{x}}$ and is given as

$$z_{\vec{x}}^{*}\ (\vec{y}) = \lambda/(\ 1 - \lambda)\ \ r(\vec{y},\vec{x}),$$

and where $\psi_{\vec{x}}^{'}\ (z_{\vec{x}}^{*})$ is the Frechêt derivative of the mapping $\psi_{\vec{x}}$ at this fixed point.

From the definition of the mapping $\psi_{\vec{x}}$ it follows however immediately that

$$\psi_{\vec{x}}^{'}\ (z_{\vec{x}}^{*}) = \lambda\,1 \qquad\qquad (III.64)$$

independent of $\vec{x}$, where 1 is the identity operator in the space $\mathcal{C}(F)$.

The operator $\psi_{\vec{x}}'(z_{\vec{x}}^*)$ therefore is not of trace class and the determinant of $1 - \psi_{\vec{x}}'(z_{\vec{x}}^*)$ cannot be defined in a reasonable way. But this shows that the formal expression (III.63) makes no sense and the operator $\mathcal{L}$ has therefore no trace. An interesting feature with the expression (III.63) is however that it gives, if one forgets about the factor $\det(1-\psi_{\vec{x}}'(z_{\vec{x}}^*))$, exactly the partition function for one lattice site of our system with periodic boundary conditions:

$$Z_1 = \int_F d\omega \, \exp\beta\left[h(\vec{x}) + 2/(1-2)\, r(\vec{x},\vec{x})\right].$$

This is also the case for the iterates $\mathcal{L}^n$ of $\mathcal{L}$ when one forgets again an undefined factor coming also from the operator $\psi_{\vec{x}}'(.)$.
This somehow gives us the feeling that the choice (III.60) for the subspace $\mathcal{f}_\infty$ was not good enough in the sense that the multiplicity of the different eigenvalues of the operator $\mathcal{L}$ in this space is still to big. The set of eigenvalues on the other hand seems to be already exactly what we wanted it to be. So one should try to restrict the domain of definition of the operator $\mathcal{L}$ further so that it becomes finally really the transfer matrix of our continuous spin system.

For a special class of functions $r(.,.) : F \times F \to \mathbb{R}$ in the interaction Φ in (III.0) the above mentioned difficulties can be handled in an easy way. These difficulties appearently arose from the fact that the mapping w defined in (III.59) maps the space $\Omega_{\supset}$ not into a finite dimensional space but in the above example into the Banach space $\mathcal{C}(F)$. This was not so in the discrete case where only finite dimensional regions in complex spaces were used.

If the function r can be written as

$$r(\vec{x},\vec{y}) = \sum_{i=1}^{m} s_i(\vec{x})\, t_i(\vec{y}) \quad , \tag{III.65}$$

with s_i and t_i continuous functions on F and $m < \infty$, the configuration space $\Omega_>$ can be mapped again into a finite dimensional space and the above mentioned difficulties do not show up.

One considers instead of (III.59) the mapping $w: \Omega_> \to \mathbb{C}^m$ defined as

$$(w(\zeta_>))_j := \sum_{i=1}^{\infty} t_j(\zeta_i) \; \lambda^i \;, \qquad\qquad 1 \le j \le m \;.$$

Then one can argue exactly as in section III.1. so that we can omit here all the details.

Examples of interactions with functions r as above are for instance the N-vector models [56] which we mentioned already previously.

A discussion of a continuous system on the real axis with an exponentially decreasing interaction has been given in [92].

In the next section we want to study another class of interactions which besides the exponentially decreasing ones considered in this section can be treated by the ideas developped in the preceding chapters. It turns out that as in the case of a general continuous spin system the Ruelle-Araki operator is again a linear operator acting in a Banach space of holomorphic functions over infinite dimensional domains. Contrary to the case discussed just above the Ruelle-Araki operator is again a trace class operator.

III.2. <u>Exponentially decreasing interactions of the form $J(i) =$</u>
<u>$= a(i) \exp{-\chi i^{\beta_1}}, \; \beta_1 > 1$</u>

In this section we make another step towards a discussion of polynomially decreasing interactions, even if at a first glance the interactions considered here decrease even faster than those treated in the last section. The important fact in the above interaction for us is however the function $a(i)$. This can be any function on the positive integers with real values with the following behavior at infinity:

$$\lim_{k \to \infty} |a(k)| \exp{-\varrho k^{\varepsilon}} = 0 \quad \text{for all} \quad \varrho, \varepsilon > 0. \tag{III.66}$$

This just says that the function a(i) can increase at infinity at most as a polynomial in i. Otherwise this function is completely arbitrary. As a special case one can certainly take the function $a(i) = i^{-m}$, $m \in \mathbb{N}$. But then one would get really polynomial decrease at infinity from the above interactions when one considers only the limit $\beta_1 \to 0$ or $\gamma \to 0$. Because we have to limit our discussion here to values of $\beta_1 > 1$ the limit $\beta_1 \to 0$ is at the moment completely out of the possibilities of our methods . The limit $\gamma \to 0$ however , which would lead also to slowly decreasing interactions, is in the range of applicability of our method. We will come back to this problem later once more.

It turns out that the case $\beta_1 = 1$ which we discussed in the fore-going section for very special functions a(i) plays a somehow special role in the sense that a treatment along the lines we will present here for $\beta_1 > 1$ is not possible for the former case. To include a treatment also of functions $J(i) = a(i) \exp{-\gamma i}$ with an arbitrary function a(i) analogous to (III.63) one would need a detailed knowled-ge of the limiting behavior of the Ruelle-Araki transfer operator in the limit $\beta_1 \to 1$ where it is not anymore a trace class operator in the spaces we consider here.

It was shown already by Ruelle[47] that the free energy of a lattice spin system with interactions of the form $J(i) = a(i) \exp{-\gamma i^{\beta_1}}$, $\beta_1 \geq 1$, is an analytic function in all relevant parameters of the system. He got this result by showing that the highest eigenvalue λ_1 of the operator $\mathcal{L}$ in the space $\ell(\Omega_{>})$ is strictly separated from the rest of the spectrum $\sigma(\mathcal{L})$ which in fact is contained in a disc of radius strictly smaller than λ_1 . His method however did not allow him to make any statements about this rest of the spectrum $\sigma(\mathcal{L}) \setminus \lambda_1$.

Because our final aim as explained above will be a discussion of the limits $\gamma \to 0$ or $\beta_1 \to 0$ it is clear that we must first have a detailed knowledge of the whole spectrum of the operator $\mathcal{L}$. This will certainly not be possible for the operator $\mathcal{L}$ in the large space $\ell(\Omega_>)$, but we hope that our method of reducing the domain of definition of $\mathcal{L}$ to certain subspaces should make such a program finally possible.

Interactions of the type $J(i) = a(i) \exp-\gamma i^{\beta_1}$ with $0 < \beta_1 < 1$ studied by Gallavotti and Lin [42] also are not tractable up to now by the method we will present here. The reason is analogous to the one for the case $\beta_1 = 1$ with arbitrary function $a(i)$ so that one gets the impression that as soon as we succeed in treating the latter case also the former can be understood without much **further efforts** .

Lets come back now to the case $\beta_1 > 1$. The interaction Φ has then in the case of a discrete spin system on a lattice the following general form

$$
\Phi(\xi_\Lambda) = \begin{cases} -h(\xi_i) & \text{for } \xi_\Lambda = (\xi_i), \Lambda = \{i\}, \\ -r(\xi_i, \xi_j)\, a(|j-i|)\, 2^{|j-i|^{\beta_1}} & \text{for } \xi_\Lambda = (\xi_i, \xi_j), \Lambda = \{i,j\}, \\ 0 & \text{otherwise,} \end{cases}
$$

where h and r are again real valued functions on the space F respectively F x F .

The partition function Z_n for the interval $\Lambda_n = [1,n]$ in $Z_>$ with periodic boundary conditions $\xi_{n+i} = \xi_i$ for all $i \in \mathbb{N}$ reads for this system as follows:

$$
Z_n = \sum_{\xi_{\Lambda_n} \in \Omega_{\Lambda_n}} \exp \beta \left[\sum_{i=1}^{n} h(\xi_i) + \sum_{i=1}^{n} \sum_{l=i+1}^{\infty} r(\xi_i, \xi_l)\, a(l-i)\, 2^{(l-i)^{\beta_1}} \right]
$$

(III.67)

with $\xi_{\Lambda_n} = (\xi_i)_{1 \le i \le n}$ and $\xi_{n+i} = \xi_i$ for all $i \in \mathbb{N}$.

The Ruelle-Araki transfer matrix $\mathcal{L}: \mathcal{C}(\Omega_{>}) \longrightarrow \mathcal{C}(\Omega_{>})$ as defined quite generally in (II.5) is then given as

$$\mathcal{L}f(\xi_{>}) = \sum_{\sigma \in F} f(\sigma, \xi_{>}) \exp\beta\left[h(\sigma) + \sum_{i=1}^{\infty} r(\sigma, \xi_i) a(i) \lambda^{i^{\beta_1}}\right], \text{(III.68)}$$

with the configuration $\xi_{>} = (\xi_i)_{i \in \mathbb{N}}$.

The procedure is now again as in all the other cases discussed in the previous sections only the mathematics gets a little bit more involved.

We assume that the set F consists of d elements which we denote again by $\sigma_1, \ldots, \sigma_d$. To describe the subspace $\mathcal{C}_{\infty}$ of $\mathcal{C}(\Omega_{>})$ in which the operator $\mathcal{L}$ above should become a trace class operator we need some notations.

Let $l_1 := \left\{ z = (z_i)_{i \in \mathbb{N}} : \|z\| = \sum_{i=1}^{\infty} |z_i| < \infty, z_i \in \mathbb{C} \right\}$ be the Banach space of all absolutely summable sequences of complex numbers. Denote then by l_1^d the d-fold direct product of d copies of this space which again is a Banach space. Let B_R be an open polyball of radius R in the space l_1^d that means

$$B_R = \left\{ \underline{z} \in l_1^d : \|z^{(i)}\| < R \quad \text{for all } 1 \leq i \leq d \right\}, \tag{III.69}$$

where we denoted the elements of the space l_1^d by $\underline{z} = (z^{(i)})_{1 \leq i \leq d}$. Furthermore we denote by $A_{\infty}(B_R)$ the Banach space of all functions f holomorphic in B_R and bounded in $\bar{B}_R$. We construct then a mapping $\underline{w}: \Omega_{>} \to l_1^d$, $\underline{w} = (w_i)_{1 \leq i \leq d}$ as

$$(w_i(\xi_{>}))_k = \sum_{j=1}^{\infty} r(\sigma_i, \xi_j) a(j+k-1) \exp-\gamma\left[(j+k-1)^{\beta_1} - (k-1)^{\beta_1}\right], \tag{III.70}$$

for $1 \leq k \leq \infty$.

Since by assumption $\beta_1 > 1$ we have for all $1 \leq i \leq d$

$$\| w_i (\xi_{>}) \| = \sum_{k=1}^{\infty} | (w_i (\xi_{>}))_k | < \infty , \qquad \text{(III.71)}$$

which means that for every $\xi_{>} \in \Omega_{>}$ the sequence $w_i (\xi_{>})$ belongs to the space l_1 .

One has even for every $\delta > 0$

$$\| w_i (\xi_{>}) \|_{\delta} \; : = \; \sum_{k=1}^{\infty} | (w_i (\xi_{>}))_k |^{\delta} < \infty .$$

Let $\sigma_0 \; : = \; \max_{\sigma_i, \sigma_j \in F} | r (\sigma_i , \sigma_j) |$ and chose a number R such that for

all $\xi_{>} \in \Omega_{>}$ and all $1 \le i \le d$ the inequality $\| w_i (\xi_{>}) \| < R$ is valid.

This can be achieved for example with any R such that

$$R > \sigma_0 \sum_{k=1}^{\infty} \sum_{i=1}^{\infty} | a(i+k-1) | \; \exp -\gamma \left[(i+k-1)^{\beta_1} - (k-1)^{\beta_1} \right] . \qquad \text{(III.72)}$$

Relation (III.71) shows that the mapping $\underline{w}$ defined in (III.70) maps the space $\Omega_{>}$ in a continuous way into the polyball B_R if R is chosen as in (III.72). Now we can define the subspace $\mathcal{C}_{\infty}$ as

$$f \in \mathcal{C}_{\infty} \Longleftrightarrow \exists g \in A_{\infty}(B_R) \; : \; f(\xi_{>}) = g(\underline{w}(\xi_{>})) \; \text{for all} \; \xi_{>} \in \Omega_{>}. \qquad \text{(III.73)}$$

To show invariance of this space under the action of the operator $\mathcal{L}$ in (III.68) we define for any $g \in A_{\infty}(B_R)$ a new function $\mathcal{L}g$ as

$$\mathcal{L}g(\underline{z}) \; : = \; \sum_{i=1}^{d} g(\psi_i (\underline{z})) \; \exp \beta \left[h(\sigma_i) + z_1^{(i)} \right] , \qquad \text{(III.74)}$$

where $\psi_i \; : \; B_R \to B_R$ is the following mapping

$$(\psi_i^{(j)} (\underline{z}))_k = \exp -\gamma \; (k^{\beta_1} - (k-1)^{\beta_1}) \; (z_{k+1}^{(j)} + a(k) \; r(\sigma_i , \sigma_j)). \qquad \text{(III.75)}$$

To make sure that ψ_i maps B_R strictly inside itself one can chose R for instance such that

$$R > \sigma_0 \ (1 - \exp{-\gamma})^{-1} \sum_{k=1}^{\infty} |a(k)| \ \exp{-\gamma} \ (k^{\beta_1} - (k-1)^{\beta_1}) \ . \qquad \text{(III.76)}$$

Finally we assume R to satisfy both the relations (III.76) and (III.72).

Since the mappings ψ_i are obviously holomorphic also the function $\mathcal{L}g$ in (III.74) is a holomorphic function in B_R and therefore an element in the space $A_{\infty}(B_R)$. A simple algebraic calculation then shows that

$$\mathcal{L} f(\underline{\xi}_{>}) = \mathcal{L}g(\underline{w}(\underline{\xi}_{>})) \ . \qquad \text{(III.77)}$$

The operator $\mathcal{L}$ in the space $A_{\infty}(B_R)$ is again of a form like the transfer operators in all the other cases we have discussed up to now. The difference is only that here the function g is not any more defined over a finite dimensional space but over the infinite dimensional space l_1^d. A similar situation we found already in our discussion of a general continuous spin system on a lattice where this gave rise to serious mathematical problems which we could not solve. In particular Theorem III.1 can no longer be applied to such a situation. As we showed however in [117] this theorem can be generalized in certain cases also to the infinite dimensional case. Here we need only a special version of what we showed in greater generality in [117] . This allows us however also to make much stronger statements than in the general case.

Consider namely the mappings $\underline{\psi}_j$ in (III.75) a little bit closer. They can be written as

$$\underline{\psi}_j(\underline{z}) = \underline{\psi}_0(\underline{z}) + \underline{z}_{o,j} \qquad \text{(III.78)}$$

where $z_{o,j} \in B_R$ is defined as

$$(z_{o,j}^{(i)})_k := r(\bar{\sigma}_j, \bar{\sigma}_i) \ a(k) \ \exp{-\gamma} \ (k^{\beta_1} - (k-1)^{\beta_1}), \qquad \text{(III.79)}$$

and $\psi_o : l_1^d \rightarrow l_1^d$ denotes the linear operator

$$(\psi_o^{(i)}(z))_k := \exp{-\gamma} \ (k^{\beta_1} - (k-1)^{\beta_1}) \ z_{k+1}^{(i)}. \qquad \text{(III.80)}$$

This linear mapping ψ_o is the direct product of the mappings $\psi_o : l_1 \rightarrow l_1$ defined as

$$\psi_o = \sum_{k=1}^{\infty} \exp{-\gamma} \ (k^{\beta_1} - (k-1)^{\beta_1}) \ e_{k+1}^* \otimes e_k. \qquad \text{(III.81)}$$

Thereby the quantities $e_k \in l_1$ and $e_{k+1}^* \in l_1^*$ are defined as

$$(e_k)_i = \delta_{ik} \quad \text{and} \quad e_{k+1}^* (z) = z_{k+1} \quad \text{for all } z \in l_1.$$

Since

$$\sum_{k=1}^{\infty} \exp{-\delta} \gamma \ (k^{\beta_1} - (k-1)^{\beta_1}) < \infty \qquad \text{for all } \delta > 0,$$

the operator ψ_o defines a nuclear operator of order zero in the Banach space l_1. But then trivially also the operator ψ_o is a nuclear operator of order zero in the space l_1^d.

In Appendix B we will prove the following theorem:

<u>Theorem III.2</u> Let B_R be an open polyball in the space l_1^d. Let $z_o \in B_R$ and let $\psi_o : l_1^d \rightarrow l_1^d$ be a linear bounded mapping such that the map $\psi(z) := z_o + \psi_o(z)$ maps B_R strictly inside itself. Let ψ_o be nuclear of order zero with $\|\psi_o\| < 1$. Let further $\varphi \in A_\infty(B_R)$.

Define the composition operator $T : A_\infty(B_R) \longrightarrow A_\infty(B_R)$ as

$$T f (\underline{z}) := \varphi(\underline{z}) \quad f \circ \psi(\underline{z}) \ .$$

Then we have

1) ψ has exactly one fixed point $\underline{z}^*$ in B_R ,

2) the operator T is also nuclear of order zero,

3) the trace of T is given by the formula

$$\text{trace } T = \varphi(\underline{z}^*) \quad \det(1 - \psi_0)^{-1},$$

where $\underline{z}^*$ is the unique fixed point of ψ .

It is now obvious that our operator $\mathcal{L}$ in (III.74) fulfills all the assumptions of this theorem. Therefore we get

Corollary III.2   The operator $\mathcal{L}$ and all its iterates $\mathcal{L}^n$ are nuclear of order zero and their traces are given by the formula

$$\text{trace } \mathcal{L}^n = \det (1 - \psi_0^n)^{-1} \ z_n,$$

where z_n is just the partition function as given in (III.67).

To show this one has only to apply the trace formula of Theorem III.2 and determine the fixed points of the mappings ψ^n arising in the expressions for the operators $\mathcal{L}^n$.

Since one can apply also in this case the theory of positive operators in Banach spaces the free energy $f(\beta)$ of our system can again be expressed as

$$- \beta f(\beta) = \log \lambda_1, \tag{III.82}$$

where λ_1 denotes as usual the highest eigenvalue of the operator $\mathcal{L}$ which is again analytic in all the parameters describing the system.

This way we got a new proof of the result of Ruelle on the analyticity of the free energy of such a one-dimensional lattice spin system with an interaction decreasing faster than exponentially at infinity. Our proof however gives a much stronger characterization of the operator $\mathcal{L}$ which when restricted to the space $A_\infty(B_R)$ is of trace class and has therefore a very nice discrete spectrum. In this space $\mathcal{L}$ is the true transfer matrix of such systems and can be compared with the original transfer matrix of Kramers and Wannier in the case of finite range interactions. In principle one can therefore express also all the correlation functions in terms of the set of eigenvalues of this operator $\mathcal{L}$ in $A_\infty(B_R)$. This kind of results is strictly impossible to get with the method of Ruelle who studied the operator $\mathcal{L}$ always in the space $\ell(\Omega)$ where the spectrum is certainly not so simple to understand.

A further application of these results will be a nice improvement of certain analyticity properties of the so called zeta-functions which one can define for such lattice systems and which we will discuss in the last chapter of this work.

Before coming to this discussion we want to add some remarks on the problem of polynomially decreasing interactions which certainly must be of a very vague and speculative nature because of our lack of understanding such systems.

III.3. Polynomially decreasing interactions

As we mentioned already in the last section one can get these interactions quite formally from the ones treated there by a limit process: if one lets in the interaction Φ with

$$J(i) = a(i) \exp{-\gamma \, i^{\beta_1}} \, ,$$

(III.83)

the parameters γ or β_1 tend to zero one gets an interaction ϕ with $J(i) = a(i)$ where $a(i)$ can be especially also the function $a(i) = i^{-m}$, with $m \in \mathbb{N}$, which is a polynomially decreasing interaction. Unfortunately we do not know of any method which would allow us to draw immediately some conclusions about the spectrum of the operator $\mathcal{L}$ corresponding to the above limits, which will be no longer of trace class. So one has to study the spectrum of trace class operators in a limit where one leaves this class of operators which is without doubt no easy job to do.

Of special interest would be certainly the limit $\beta_1 \to 0$, because there one has the first problem already in going from $\beta_1 > 1$ to the limit $\beta_1 = 1$ and also to the region $0 < \beta_1 < 1$. Here one could hope really to find a way to apply our methods with not to much efforts because the interactions decrease nevertheless still very fast at infinity.

How could this goal possibly be achieved? We have found in the case $\beta_1 > 1$ that the partition functions Z_n are given up to certain factors by the traces of the operators $\mathcal{L}^n$. These functions are as one can convince oneself very easily holomorphic functions in the variable β_1 in the halfspace $\mathrm{Re}\,\beta_1 > 1$. The problem now is, if these functions can perhaps be analytically continued also outside this region, especially also onto the real axis. One could hope that this continuation somehow can be done than also on the level of the operator $\mathcal{L}$ and that one arrives this way at some new trace class operator which describes such a system.

A procedure a little bit different from the one described just now to treat polynomially decreasing interactions and which we want to discuss next leads to the kind of problem we encountered already in the discussion of continuous spin systems in section III.1.3.

Instead of an interaction with $J(i) = i^{-m}$ one considers another model first introduced by M. Kac [55] with an interaction ϕ with

$J(i)$ given as $J(i) = J_0 \int_0^1 \tau^\alpha \exp{-\gamma\tau i}\ d\tau$. For large i this function $J(i)$ behaves like $J(i) \sim i^{-\alpha+1}$.

One can write down the Ruelle-Araki operator for such an interaction
which reads in the special case of an Ising system with vanishing exterior field h

$$\mathcal{L} f(\xi_>) = \sum_{\sigma \in F} f(\sigma,\xi_>)\ \exp\left[\beta J_0 \sigma \sum_{i=1}^\infty \xi_i \int_0^1 \tau^\alpha \exp{-\gamma\tau i}\ d\tau\right]. \qquad (III.84)$$

Analogous to our previous procedure we map the configuration space
into another Banach space which in this case can be taken as the space
$\mathcal{C}\left[(0,1)\right]$ of all continuous complex valued functions over the closed
interval $[0,1]$ in $\mathbb{R}$:

$$w(\xi_>)(\tau) := \tau^\alpha \sum_{i=1}^\infty \xi_i\ \exp{-\gamma\tau i} . \qquad (III.85)$$

To guarantee $w(\xi_>)$ really to belong to the space $\mathcal{C}([0,1])$ we have to
restrict α to values $\alpha \geq 1$. Consider then again the space $A_\infty(B_R)$ of
holomorphic functions over an appropriately chosen ball B_R in the
Banach space $\mathcal{C}([0,1])$. In the space $A_\infty(B_R)$ the Ruelle-Araki operator $\mathcal{L}$ is given as

$$\mathcal{L} g(z) = \sum_{\sigma \in F} g(\psi_\sigma(z))\ \exp\left[\beta\sigma J_0 \int_0^1 z(\tau)\ d\tau\right], \qquad (III.86)$$

where $\psi_\sigma : B_R \to B_R$ denotes this time the following holomorphic mapping
for $\sigma \in F$:

$$(\psi_\sigma(z))(\tau) := \sigma \tau^\alpha \exp{-\gamma\tau} + \exp{-\gamma\tau}\ z(\tau) .$$

This way the operator $\mathcal{L}$ is a well defined linear operator in the space
$A_\infty(B_R)$ when R is chosen in the right way. It turns out however, that
also this operator like the one defined in (III.61) for the continuous

spin system is not of trace class. This follows again from the fact that the linear operators $\mathcal{Y}_{\sigma}^{0} : \mathcal{L}([0,1]) \longrightarrow \mathcal{L}([0,1])$

$$(\mathcal{Y}_{\sigma}^{0}(z))(\tau) := \exp{-\int \tau} \, z(\tau)$$

arising as the Frechêt derivative of the mappings $\mathcal{Y}_{\sigma}$ above are not of trace class. And it is exactly again the factor coming from det $(1 - \mathcal{Y}_{\sigma}^{0})$ which destroys the applicability of the operator $\mathcal{L}$ as defined in (II.86). The other term appearing in the trace formula when formally applied gives again the correct partition function.

We interpret this as some sign that our procedure outlined above is somehow not yet fully developped to overcome these difficulties. It would be interesting to find quite generally methods which enable us to find for the above operators those restrictions of the domain of definition which make such operators then to trace class operators. It is certainly clear from the above discussion that our choice for the models discussed in this section is not yet the right restriction .

It is also possible that such polynomially decreasing interactions are completely out of the range of applicabilty of the ideas presented in this work. But this is an open problem to which we can not give an answer at the moment.

IV. ZETA-FUNCTIONS OF CLASSICAL ONE-DIMENSIONAL SYSTEMS

IV.1. Definitions and general properties

Zeta-functions are fascinating objects in mathematics and have been introduced in statistical mechanics only recently by Ruelle [106], [118]. These functions are natural generalisations of the zeta-function discussed first by Artin and Mazur [119] in the theory of dynamical systems. These authors showed how these functions describe in a very compact way some global properties of such dynamical systems. A dynamical system in its abstract definition is quite generally a topological space M together with a continuous mapping $f: M \to M$ on it. In most cases of interest the space M has also a differentiable structure and f is a diffeomorphism. Among these dynamical systems one finds all classical systems of mechanics which are given by ordinary differential equations in the phase space of the system. As we saw however not all dynamical systems in physics are of this kind. The configuration space Ω of classical lattice spin systems is not a smooth manifold but only a discrete metrizable compact space. This is the reason why we have to consider here the more general scheme of topological dynamical systems. Artin and Mazur found that in a smooth dynamical system the transformation $f : M \to M$ can be well characterized by its phase portrait that means the orbit structure like fixed points and closed orbits. Such closed orbits are nothing else but again fixed points of the iterates $f^n : = f \circ \cdots \circ f$ of the mapping f. As was shown in [21] these fixed points of all the mappings f^n determine also the statistical behavior of such a dynamical system (M, f).

Let then

$$N_n(f) : = \# \left\{ x \in M: f^n(x) = x \right\} \tag{IV.1}$$

One tries to combine all the information about these numbers in one function. This can be done as follows:

Consider the formal series

$$\mathcal{S}(z) := \exp \sum_{n=1}^{\infty} z^n/n \; N_n(f) \; .$$

(IV.2)

Artin and Mazur [119] could prove that this formal expression makes indeed sense for almost all diffeomorphisms of a compact manifold M. They showed that the above series has a non-vanishing radius of convergence and defines therefore a holomorphic function in some small neighbourhood of $z = 0$ in $\mathbb{C}$. One can interpret their result also as showing that the numbers N_n of fixed points of the mappings f^n grow at most like c^n, where c is some finite positive constant.

A detailed discussion of the known analyticity properties of the function $\mathcal{S}(z)$ for so called Axiom A respectively Anosov systems can be found in Smale [25] .

To understand the importance of this function $\mathcal{S}(z)$ for the statistical mechanics of classical one-dimensional systems we consider our spin system on the positive axis $Z_>$ with the corresponding configuration space $\Omega_>$. On this space we had defined the shift operator :

$$(\tau \xi_>)_i = \xi_{i+1} \quad \text{for all } i \in \mathbb{N} \; .$$

This means τ transforms the space $\Omega_>$ into itself by shifting any configuration $\xi_>$ one lattice site to the left. The pair $(\Omega_>, \tau)$ defines therefore a topological dynamical system in the sense defined above. It is known in the mathematical literature under the name of a one-sided subshift of finite type d, if d is the number of elements of the set F. In ergodic theory it is also called a Bernoulli shift of order d [79], [120] .

For this simple dynamical system the function $\mathcal{S}$ defined above can be calculated without problems. It is obvious that a configuration $\xi_>$

belongs to the fixed point set Fix τ^n of the shift operator τ if, and only if $\xi_{i+n} = \xi_i$ for all $i \in \mathbb{N}$ that means $\xi_>$ is a periodic configuration of period n . This shows that $N_n(\tau) = d^n$ and the function $\zeta(z)$ in (IV.2) can be calculated to be

$$\zeta(z) = \exp \sum_{n=1}^{\infty} z^n/n \; d^n = \exp-(\log(1 - dz)) = (1 - dz)^{-1} . \qquad (IV.3)$$

This calculation is certainly true for small enough z but by the procedure of analytic continuation we see that the zeta-function of a free spin system on a lattice is a meromorphic function in the whole z plane. The function has a simple pole at $z = d^{-1}$.

On the other hand we know the free energy of such a simple system without interaction :

$$-\beta \, f(\beta) = \log d . \qquad (IV.4)$$

Using this we can interpret the location of the simple pole in the corresponding zeta-function as given in (IV.3) in a more physical way: the zeta-function has a pole at

$$z = \exp \beta \, f(\beta) .$$

It is clear that this free system is not of much interest from a physical point of view and also the location of the pole in the zeta function seems to be not very exciting . We will see however that there is a much deeper connection between this function and the physical properties of such one-dimensional lattice spin systems.

First of all one has to generalize these functions in such a way that they can describe also physically more interesting situations as for instance systems with non-vanishing interactions. This was accomplished by Ruelle[118]. Ruelle's idea thereby was rather simple and

from the statistical mechanics point of view straightforward: in the
Artin- Mazur function as defined in (IV.2) all fixed points $x \in \text{Fix } f^n$
are counted with the same weight as equal important, that means translated into the language of spin systems, all configurations in configuration space $\Omega_>$ are equal probable. In the case of a system without
interactions this is certainly true. As soon as there are however non-
vanishing interactions we know that one has to take nontrivial measures
in configuration space to get out physical results. These measures we
discussed in the first chapter of this work. They attribute to different configurations $\xi_>$ different weights which are determined by the
energy function $U(\xi_>)$ as defined in (I.11). Having this in mind one
can understand Ruelle's generalized zeta-functions for dynamical systems
without difficulties. Their exact definition is as follows:

Let M be a topological space and $f: M \to M$ a continuous mapping.
Let $A : M \to \mathbb{C}$ be any complex valued function on M which in the case of
our one-dimensional spin system will describe something like the above
mentioned energy of a configuration. Instead of (IV.2) one considers
the following formal expression

$$\int (z,A) : = \exp\left[\sum_{n=1}^{\infty} z^n/n \sum_{x \in \text{Fix } f^n} (\exp \sum_{k=0}^{n-1} A(f^k x))\right]. \qquad (IV.5)$$

Interesting properties of this function for certain dynamical systems
have been discussed by Ruelle in [106] and [118]. Here, however, we will
restrict our attention exclusively to the case of one-dimensional spin-
systems on a lattice and the corresponding dynamical system $(\Omega_>, \tau)$.

Let ϕ be a two-body interaction as in (I.24) resp. (I.25). Define
next a function $A : \Omega_> \to \mathbb{C}$ as

$$A(\xi_>) : = -\beta \left(\sum_{j=2}^{\infty} \phi_2(\xi_1, \xi_j) + h(\xi_1)\right), \qquad (IV.6)$$

where the configuration $\xi_>$ is given as $\xi_> = (\xi_i)_{i \in \mathbb{N}}$. This function

A is continuous on Ω, for the interaction ϕ as chosen above.

We are going to calculate Ruelle's zeta-function for this special choice of A. We know already that $\xi_> \in \mathrm{Fix}\,\tau^n$ if, and only if $\xi_>$ is periodic with period n, that is $\xi_{i+n} = \xi_i$ for all $i \in \mathbb{N}$.

Consider then the expression

$$U_n(\xi_>) = \sum_{k=0}^{n-1} A(\tau^k \xi_>) .$$

Using (IV.6) we can write this as

$$U_n(\xi_>) = -\beta \sum_{k=0}^{n-1} (\sum_{j=k+2}^{\infty} \phi_2(\xi_{k+1}, \xi_j) + h(\xi_{k+1})) =$$

$$= -\beta (\sum_{k=1}^{n} \sum_{j=k+1}^{\infty} \phi_2(\xi_k, \xi_j) + \sum_{k=1}^{n} h(\xi_k)) . \qquad (IV.7)$$

But this is just the contribution of the configuration $\xi_>$ to the entire energy of our system with periodic boundary conditions. This then shows that

$$\sum_{\xi_> \in \mathrm{Fix}\,\tau^n} \exp \sum_{k=0}^{n-1} A(\tau^k \xi_>) = Z_n , \qquad (IV.8)$$

where Z_n denotes the partition function of the one-dimensional system for n lattice sites with periodic boundary conditions. Therefore Ruelle's zeta-function reads in this special case of the dynamical system $(\Omega_>, \tau)$ with the function A as in (IV.6) as

$$\zeta(z,A) = \exp \sum_{n=1}^{\infty} z^n/n \; Z_n . \qquad (IV.9)$$

We can determine without difficulties the radius of convergence ρ of this function. According to the formula of Cauchy and Hadamard [115] one has

$$\rho^{-1} = \lim_{n \to \infty} \sup |Z_n|^{1/n} . \qquad (IV.10)$$

Since by definition of the free energy $f(\beta)$

$$\exp - \beta \, f(\beta) = \lim_{n \to \infty} z_n^{1/n} \, ,$$

it follows that

$$\varrho = \exp \beta \, f(\beta) \, . \tag{IV.11}$$

We can then apply a theorem of Pringsheim [121] according to which a power series with positive coefficients has its first singularity on the real line in the point $z = \varrho$, where ϱ is just the radius of convergence of this series. Therefore the series in (IV.9) defines a holomorphic function in the disc D_R where $R = \exp \beta \, f(\beta)$. Furthermore this function has a singularity at the point $z = \exp \beta \, f(\beta)$. This is in agreement with the result for the Artin-Mazur function in the free case to which the Ruelle function is reduced for vanishing A or, what is the same, vanishing interaction Φ.

The above reasoning gives therefore a simple explanation for this fact. But this shows that the function $\zeta(z,A)$ determines via its analyticity properties, especially through the position of its first pole on the real axis, the physical properties of such a lattice system which are completely determined by the free energy $f(\beta)$. From this it should be clear that this function is an interesting object also from the physical point of view.

We will see how one can get even stronger results for these functions when the interaction Φ is more restricted in its behavior at infinity. At the same time we want to generalize our systems a little bit further in as far as we want to include in our treatment also systems with certain restrictions on the allowed configurations as we considered them already in the first chapter of this work. These restrictions will be of the following kind: the spin variable σ cannot have arbitrary values on neighbouring lattice sites, some should be allowed, others should be forbidden. A generalization to restrictions on the allowed spin values on lattice sites an arbitrary finite distance apart can be carried out in complete analogy to our following procedure.

To describe such restrictions formally let $\mathbb{T}$ be a $d \times d$ matrix indexed by the elements of the set F of possible spin values whose matrix elements are either zero or one. We then call a configuration $\xi_>$, $\xi_> = (\xi_i)_{i \in \mathbb{N}}$ an allowed configuration if for all $i \in \mathbb{N}$ the matrix elements $\mathbb{T}_{\xi_i, \xi_{i+1}} = 1$. It is obvious that the shift operator τ in (I.4) maps allowed configurations onto allowed configurations.

The triple $(\Omega_>, \tau, \mathbb{T})$ is called a one-sided subshift of finite type with transition matrix $\mathbb{T}$ [19].

The Artin-Mazur zeta-function for the free one-sided subshift of finite type with transition matrix $\mathbb{T}$ was determined by Bowen and Lanford [122] and is given by the expression

$$\zeta(z) = \det(1 - z\mathbb{T})^{-1} .$$ (IV.12)

For the case $\mathbb{T}_{\sigma_i, \sigma_j} = 1$ for all $\sigma_i, \sigma_j \in F$ one recovers just expression (IV.3).

In the free case the Artin-Mazur function is therefore even a rational function and meromorphic in the entire z plane.

We will show that a similar result is true also for Ruelle's zeta-function if the range of the interaction Φ is either finite or the interaction vanishes exponentially fast at infinity. These results we obtain again with the help of the Ruelle-Araki transfer matrix formalism which we developped in the former chapters. To include the restrictions on the allowed configurations as described by the transition matrix $\mathbb{T}$ we have to modify a little bit the definition of this Ruelle-Araki operator as given in (II.5).

To do this let $\mathcal{C}(F)$ be the space of continuous functions on F which is a finite dimensional vector space. Define next the Banach space $B = \mathcal{L}(F) \hat{\otimes}_{\pi} \mathcal{L}(\Omega_{>})$ as the topological projective tensor product of these two spaces together with the π-norm (see Appendix A for definitions). The transfer matrix is then defined as

$$\mathcal{L} f(x,\xi_{>}) : = \sum_{\sigma \in F} T_{\sigma}(x) \ f(\sigma,(\sigma,\xi_{>})) \ \exp{-\beta\left[h(\sigma) + \sum_{j=1}^{\infty} \phi(\sigma,\xi_{j})\right]}, \quad (IV.13)$$

with $f = f(x,\xi_{>}) \in B$ and x an arbitrary element in F. The function $T_{\sigma} \in \mathcal{L}(F)$ is thereby defined as

$$T_{\sigma}(x) : = \mathbb{T}_{\sigma,x} . \qquad\qquad (IV.14)$$

It is easy to see that $\mathcal{L}$ defines a bounded linear operator in the space $B = \mathcal{L}(F) \hat{\otimes}_{\pi} \mathcal{L}(\Omega_{>})$ if the interaction ϕ is again of the form given in (I.25). Also Theorem II.2 of chapter II. can be generalised immediately to this case. Since we consider here only interactions with finite range respectively those with exponential decrease at infinity we will not persue the general discussion any further and restrict ourselves to these last mentioned cases.

IV.2. Ruelle's zeta-function for finite range interactions

We consider once more interactions of the form given in (I.27). From definition (IV.13) of the operator $\mathcal{L}$ one sees that it can be written as an operator in the space $B = \mathcal{L}(F) \hat{\otimes}_{\pi} \mathcal{L}(\Omega_{>})$ as follows:

$$\mathcal{L} = \sum_{\sigma \in F} 1_{\sigma} \otimes t_{\sigma} . \qquad\qquad (IV.15)$$

Thereby $1_{\sigma} : \mathcal{L}(F) \longrightarrow \mathcal{L}(F)$ is the finite rank operator

$$1_{\sigma} \varphi(x) : = T_{\sigma}(x) \ \varphi(\sigma) \quad \text{for} \quad \varphi \in \mathcal{L}(F), \qquad\qquad (IV.16)$$

and $t_\sigma : \mathcal{C}(\Omega_>) \longrightarrow \mathcal{C}(\Omega_>)$ denotes the following linear operator :

$$t_\sigma \, g(\xi_>) \; : \; = \; g(\sigma, \xi_>) \; \exp{-\beta \left[h(\sigma) \; + \sum_{k=1}^{\infty} \phi_2(\sigma, \xi_k) \right]} \, . \qquad (IV.17)$$

The symbol $\otimes$ in relation (IV.15) denotes the tensor product of the two linear mappings 1_σ and t_σ in the space $\mathcal{C}(F) \hat{\otimes}_\pi \mathcal{C}(\Omega_>)$.

In the discussion of systems with finite range interactions in chapter II.2.1. we saw that the operator t_σ leaves invariant the sub-space $\mathcal{C}_r$ which was defined in II.2.1.. Therefore this operator could be represented in this space by a $d^r \times d^r$ matrix $\mathbb{L}$. In the present case the operator $\mathcal{L}$ in (IV.15) leaves invariant the space $\mathcal{C}(F) \hat{\otimes}_\pi \mathcal{C}_r$ which is the ordinary tensor product $\mathcal{C}(F) \otimes \mathcal{C}_r$ and therefore also a finite dimensional vector space. But this means that the linear operator $\mathcal{L}$ has also a representation as a $d^{r+1} \times d^{r+1}$ real valued matrix $\mathbb{M}$ in this space as long as the interaction is of finite range.

But then it is again trivial to determine the trace of the opera-tor $\mathcal{L}$ when restricted to this finite dimensional vector space. Accor-ding to expression (IV.15) we get

$$\text{trace} \, \mathcal{L} \; = \sum_{\sigma \in F} \quad \text{trace} \, 1_\sigma \quad \text{trace} \, t_\sigma \, . \qquad (IV.18)$$

The trace of 1_σ however is given according to (IV.16) simply as

$$\text{trace} \, 1_\sigma \; = \; T_\sigma(\sigma) \; = \; \mathbb{T}_{\sigma, \sigma} \, .$$

The trace of the operator t_σ on the other hand has been calculated already in section II.2.1. and we got there

$$\text{trace} \, t_\sigma \; = \quad \text{contribution of the configuration } (\sigma, \sigma, \, .. \,) \text{ to the partition function for one lattice site in the system with periodic boundary conditions.}$$

From this it follows that the trace of the operator $\mathcal{L}$ in the subspace $\mathcal{C}^{(F)} \otimes \mathcal{C}_r$ is identical to the partition function of our spin system for one lattice site with periodic boundary conditions and where the restrictions on the allowed configurations are described by the matrix $\mathbb{T}$. Without giving the details for the proof for n lattice sites we can summarize the above considerations in a lemma:

Lemma IV.1 Let $(\Omega_{>}, \tau, \mathbb{T})$ be a subshift of finite type and let Φ be a finite range interaction. Then the partition function Z_n for n lattice sites with periodic boundary conditions is given as

$$Z_n = \text{trace } \mathbb{M}^n,$$

where $\mathbb{M}$ denotes the $d^{r+1} \times d^{r+1}$ matrix defined above.

An immediate consequence of this lemma for Ruelle's zeta-function is the following corollary

Corollary IV.1 Ruelle's zeta-function for the one-sided subshift of finite type $(\Omega_{>}, \tau, \mathbb{T})$ with finite range interaction is a rational function in the entire z plane.

Proof: Because of Lemma IV.1 we get

$$\zeta(z,A) = \exp \sum_{n=1}^{\infty} z^n/n \ Z_n = \exp \sum_{n=1}^{\infty} z^n/n \ \text{trace } \mathbb{M}^n =$$

$$= \exp - \left[\text{trace } \log(1 - z\,\mathbb{M})\right] = \det(1 - z\,\mathbb{M})^{-1}$$

for small enough z. Since the determinant of the matrix $(1-z\,\mathbb{M})$ is a polynomial in z the assertion of the corollary follows by analytic continuation.

IV.3. Ruelle's zeta-function for exponentially decreasing interactions

In [22] Ruelle proved the following theorem for systems with exponentially decreasing interactions:

__Theorem IV.1__ (Ruelle) There exists a number $R > \exp \beta f(\beta)$ such that
the function $\zeta(z,A)$ is meromorphic in the disc D_R of radius R in the
complex z plane and has there only one simple pole in the point $z =$
$\exp \beta f(\beta)$.

For interactions with pure exponential behavior at infinity that
means for interactions with $J(i) = P(i) \exp{-\gamma i}$, $\gamma > 0$ and P some
polynomial in i, it was shown in [103] and [105] that the above zeta-function is meromorphic even in the entire complex z plane.

Using the methods introduced in chapter II.2. we are going to
show a similar result for a larger class of interactions which we discussed already in that chapter.

__Theorem IV.2__ Let $(\Omega_\gamma, \tau, \mathbb{T})$ be a one-sided subshift of finite type
with an interaction as defined in III.2., Then Ruelle's zeta-function

$$\zeta(z,A) = \exp \sum_{n=1}^{\infty} z^n/n \; Z_n$$

has the following properties:
It is a holomorphic function in any disc of radius $R < \exp \beta f(\beta)$ and
can be analytically continued to a meromorphic function in the entire
z plane.

__Proof:__ One considers the operator $\mathcal{L}$ in (IV.13) in the Banach space
$B := \ell(F) \hat{\otimes}_\pi A_\infty(B_R)$ where $A_\infty(B_R)$ has been introduced in chapter III.2.
as the Banach space of holomorphic functions over the open polyball B_R
of radius R in the space l_1^d . Analogous to the procedure in [105] one
shows that the operator $\mathcal{L}$ can be written as

$$\mathcal{L} = \sum_{\mathfrak{G} \in F} 1_{\mathfrak{G}} \otimes t_{\mathfrak{G}} \; . \tag{IV.19}$$

Thereby $1_{\mathfrak{G}}$ is the linear finite rank operator introduced in (IV.16) and $t_{\mathfrak{G}}$ denotes the composition operator $t_{\mathfrak{G}} : A_{\infty}(B_R) \longrightarrow A_{\infty}(B_R)$ introduced in (III.74)

$$t_{\mathfrak{G}} \, f(\underline{z}) = f \circ \gamma_{\mathfrak{G}}(\underline{z}) \quad \varphi_{\mathfrak{G}}(\underline{z}) \; . \tag{IV.20}$$

The mapping $\gamma_{\mathfrak{G}}$ respectively the function $\varphi_{\mathfrak{G}}$ have been defined in (III.74) respecively (III.75). From the discussion there we know that the operator $t_{\mathfrak{G}}$ is nuclear of order zero. But then it follows from a result of Grothendieck [123] that also the tensor product $1_{\mathfrak{G}} \otimes t_{\mathfrak{G}}$ is nuclear of order zero in the projective topological tensor product B of the two Banach spaces $\mathcal{C}(F)$ and $A_{\infty}(B_R)$. This is so because $1_{\mathfrak{G}}$ is trivially nuclear of order zero as a finite rank operator. Because of relation (IV.19) the same is true then for the operator $\mathcal{L}$ in this space. Using finally the trace formula as given in Theorem B.1 of Appendix B and the formula trace $1_{\mathfrak{G}} \otimes t_{\mathfrak{G}}$ = trace $1_{\mathfrak{G}}$ trace $t_{\mathfrak{G}}$ [124], we get

<u>Lemma IV.2</u>   The partition function Z_n for n lattice sites of the one-dimensional spin system with an interaction Φ as in III.2. and restrictions on the allowed configurations described by $\mathbb{T}$ is given as

$$Z_n = \det(1 - \underline{\gamma}_0{}^n) \; \text{trace} \; \mathcal{L}^n,$$

where γ_0 is the linear operator defined in (III.80) and the operator $\mathcal{L}$ is defined in (IV.19).

This lemma allows us to calculate explicitly Ruelle's zeta-function for such systems. We find namely because of the above lemma

$$\zeta(z,A) = \exp \sum_{n=1}^{\infty} z^n/n \; \det(1 - \underline{\gamma}_0{}^n) \; \text{trace} \, \mathcal{L}^n \; .$$

Since $\underline{\gamma}_0$ is nuclear of order zero one has [95]

$$\det (1 - \underline{\gamma}_0{}^n) = \prod_k (1 - \lambda_k^n) , \qquad (IV.21)$$

where the $\{\lambda_k\}$ are the eigenvalues of the operator $\underline{\gamma}_0$ counted according to their algebraic multiplicities. The above expression for
$\det (1 - \underline{\gamma}_0{}^n)$ can be rewritten in a slightly different way. To do
this let $\underline{\alpha} = (\alpha_i)_{i \in \mathbb{N}}$ be a multiindex with $\alpha_i = 0$ or 1 for all $i \in \mathbb{N}$.
Let $|\underline{\alpha}| = \sum_{\{i\}} \alpha_i < \infty$. Then we can write

$$\det (1 - \underline{\gamma}_0{}^n) = \sum_{k=0}^{\infty} \sum_{\underline{\alpha}, |\underline{\alpha}|=k} (-1)^k \lambda^{n\underline{\alpha}} , \qquad (IV.22)$$

where $\lambda^{\underline{\alpha}} = \prod_{k=1}^{\infty} \lambda_k^{\alpha_k}$.

Using this expression for $\det (1 - \underline{\gamma}_0{}^n)$ we get for the function
$\zeta(z,A)$

$$\zeta(z,A) = \exp\left[\sum_{k=0}^{\infty} \sum_{\underline{\alpha}, |\underline{\alpha}|=k} \sum_{n=1}^{\infty} (-1)^k \text{ trace } z^n/n \mathcal{L}^n \lambda^{\underline{\alpha}n}\right] =$$

$$= \exp\left[\sum_{k=0}^{\infty} \sum_{\underline{\alpha}, |\underline{\alpha}|=k} (-1)^{k+1} \text{ trace } \log(1 - z\mathcal{L}\lambda^{\underline{\alpha}}) \right] =$$

$$= \prod_{k=0}^{\infty} \prod_{\underline{\alpha}, |\underline{\alpha}|=k} \det (1 - z\mathcal{L}\lambda^{\underline{\alpha}})^{(-1)^{k+1}} , \qquad (IV.23)$$

which is valid for z small enough. Thereby we used the formula

$$\exp \text{ trace } \log(1 - \mathcal{L}) = \det (1 - \mathcal{L}) \qquad (IV.24)$$

valid also for nuclear operators of order zero [130].

Since the Fredholm determinant $\det(1-z\mathcal{L})$ of a nuclear operator
of order zero is a holomorphic function in the entire z plane [125] the
function $\zeta(z,A)$ can be analytically continued to a meromorphic function
in the entire z plane.

That the pole at the point $z = \exp \beta f(\beta)$ is simple follows either from the theory of positive operators in a Banach space as described in Appendix C which can be without difficulties extended to the type of operators like $\mathcal{L}$. Alternatively one can take this assertion directly from Ruelle's Theorem. But this concludes the proof of Theorem IV.2 .

It is perhaps interesting to remark that Gallavotti [126] showed that Ruelle's zeta-function is in general not a meromorphic function in the entire z plane as one could conjecture. As a counter example to this conjecture he calculated this function for the Fisher-Felderhof model [127] and showed that it has in this case cuts in the z plane.

Since this model has a first order phase transition one could guess that at least systems without phase transitions have always meromorphic zeta-functions. But this seems also not to be the case, at least for continuous systems. In [114] we discussed the zeta-function for a continuous hard rod system with exponentially decreasing interaction. The corresponding zeta-function could be calculated exactly only in the case of vanishing interaction. But it turned out that it is not a meromorphic function even in the free case. This certainly supports the conjecture that the same will be true also in the case of a non-vanishing interaction.

Summarizing we see that these generalized zeta-functions introduced by Ruelle provide an interesting alternative for a mathematical rigorous description of classical one-dimensional systems of statistical mechanics. There remain certainly many open questions about these functions and their relation to the physical properties of such a system. It is for instance not quite understood how the existence of a phase transition for such a simple one-dimensional system is precisely reflected in the behavior of the corresponding zeta-function. What about the zeta-function of a system with a polynomially decreasing interaction which does not have a phase transition according to the results of Dobrushin ? Is it perhaps possible to get some results about such systems via the zeta-function which certainly then had to be calculated by different methods as those discussed in this work ?

Our method, which is based on the existence of a transfer matrix with very strong spectral behavior can be generalised to more slowly decreasing potentials only with some difficulties as we tried to explain in this work. In fact we do not know if our method works at all for polynomially decreasing interactions or if there are principal objections to such a method in these cases.

Nevertheless we regard the results of Dobrushin as a support for our following conjecture:

Conjecture: The Ruelle zeta-function is a meromorphic function in the entire z plane for a one-dimensional lattice system with polynomially decreasing interaction as long as there are no phase transitions in the system. In the case of a system with a phase transition this function will be a singular function.

We hope to come back to this conjecture in the future.

APPENDIX A.

GROTHENDIECK'S THEORY OF NUCLEAR OPERATORS IN BANACH SPACES

In this appendix we recall the fundamentals of Grothendieck's theory about nuclear operators in Banach spaces as far as they are necessary for the considerations in this work. A much more extensive presentation of this theory can be found in [91] and [128].

A.1. The projective topological tensor product of Banach spaces

Let E, F be two complex Banach spaces with norms $\| \ \|_E$ and $\| \ \|_F$. Let $E \otimes F$ be the tensor product of these two spaces with the norm $\| \ \|_\pi$ defined by

$$\|x\|_\pi : = \inf \sum_{\{i\}} \|e_i\|_E \|f_i\|_F \ , \tag{A1}$$

where the infimum has to be taken over all possible finite representations of $x \in E \otimes F$ in the form

$$x = \sum_{\{i\}} e_i \otimes f_i \ , \tag{A2}$$

with $e_i \in E$ and $f_i \in F$.

The completion of the space $E \otimes F$ under this norm is denoted by $E \hat{\otimes}_\pi F$ and is called the projective topological tensor product of the two spaces E and F. The norm defined in (A1) and introduced first by R.Schatten [129] is called the π-norm. The elements of the space $E \hat{\otimes}_\pi F$ are the Fredholm kernels.

A.2. The tensor product of two linear mappings

The following important Theorem was proved by Grothendieck [130]:

__Theorem A.1__ Let E, F, G be three Banach spaces and let $T : E \times F \longrightarrow G$

be a bilinear continuous mapping of the direct product E x F into G.
Then there exists an uniquely determined linear continuous mapping
$\hat{T} : E \hat{\otimes}_{\pi} F \longrightarrow G$ with $\hat{T} u = T (e,f)$ if $u = e \otimes f$. Furthermore one
has $\|\hat{T}\| = \|T\|$.

Consider then two pairs (E_1, F_1) and (E_2, F_2) of Banach spaces and
two linear continuous mappings $T_i : E_i \longrightarrow F_i$, $i = 1, 2$. Define a mapping $T_1 \times T_2 : E_1 \times E_2 \longrightarrow F_1 \hat{\otimes}_{\pi} F_2$ as follows:

$$T_1 \times T_2 (e_1, e_2) : = T_1(e_1) \otimes T_2(e_2) . \tag{A3}$$

According to Theorem A.1 there exists an uniquely determined linear
mapping

$$T_1 \otimes T_2 : E_1 \hat{\otimes}_{\pi} E_2 \longrightarrow F_1 \hat{\otimes}_{\pi} F_2 , \tag{A4}$$

which is called the tensor product of the two linear mappings T_1 and
T_2.

A.3. Nuclear operators in Banach spaces

Let E, F be complex Banach spaces. Let E^* be the dual Banach space
of E that means the space of all bounded linear functionals f on E with

$$\|f\| : = \sup_{\substack{e \in E \\ \|e\| \leq 1}} |f(e)| < \infty . \tag{A5}$$

Consider then the space $E^* \hat{\otimes}_{\pi} F$. Every element $L \in E^* \hat{\otimes}_{\pi} F$ defines in
a canonical way a bounded linear operator $\mathcal{L} : E \longrightarrow F$. In fact, every
L has a representation of the form

$$L = \sum_{\{i\}} \lambda_i \, e_i^* \otimes f_i \quad \text{with} \quad \sum_{\{i\}} |\lambda_i| < \infty , \tag{A6}$$

where $e_i^* \in E^*$, $f_i \in F$ all have norm one. Therefore $\mathcal{L}$ can be defined as

$$\mathcal{L} e \; : \; = \sum_{\{i\}} \lambda_i \; e_i^*(e) \; f_i \qquad \text{for } e \in E. \qquad (A7)$$

The correspondence $L \longrightarrow \mathcal{L}$ defines a mapping $\varphi \colon E^* \hat{\otimes}_\pi F \longrightarrow B(E, F)$ where $B(E, F)$ denotes the space of all linear bounded mappings of E into F. Unfortunately however it is not known if this mapping is one-to-one.

<u>Definition A.1</u> Let E, F be Banach spaces. Let $L^1(E, F) \; : \; = \varphi(E^* \hat{\otimes}_\pi F) \subset B(E, F)$. The elements of $L^1(E, F)$ are called nuclear operators or sometimes also Fredholm operators. The norm induced by π in the space $L^1(E, F)$ is the trace norm or the nuclear norm.

The space $L^1(E,F)$ is in general a quotient space of the space $E^* \hat{\otimes}_\pi F$.

A.4. The trace functional

Let E be a Banach space and E^* its dual space. Let $L \in E^* \hat{\otimes}_\pi E$ have the representation

$$L = \sum_{\{i\}} \lambda_i \; e_i^* \otimes e_i \quad , \qquad (A8)$$

with $\{\lambda_i\} \in l_1$, $e_i^* \in E^*$, $e_i \in E$, $\|e_i^*\| \leq 1$, $\|e_i\| \leq 1$.
Then consider the expression

$$\text{trace } L \; : \; = \sum_{\{i\}} \lambda_i \; e_i^*(e_i) \quad . \qquad (A9)$$

This is well defined and in fact a linear continuous functional on the space $E^* \hat{\otimes}_\pi E$.

Since it is not known in general if the mapping φ in (A7) is one-to-one it is not possible to say that a nuclear operator has a trace. Remember that this is different in the case of a Hilbert space where a nuclear operator has always a trace. To get trace class operators in a general Banach space Grothendieck introduced another class of nuclear operators which he called p-summable operators.

Let $0 < p \leq 1$ and let E, F be two Banach spaces. A Fredholm kernel $L \in E^* \hat{\otimes}_\pi F$ is called p-summable if L has the following representation

$$L = \sum_{\{i\}} \lambda_i \; e_i^* \otimes f_i \; , \tag{A10}$$

with $e_i^* \in E^*$, $f_i \in F$, $\|e_i^*\| \leq 1$, $\|f_i\| \leq 1$ and $\{\lambda_i\} \in l_p$ which means $\sum_{\{i\}} |\lambda_i|^p < \infty$.

A linear nuclear operator $\mathcal{L}: E \rightarrow F$ is called p-summable if there exists a p-summable Fredkolm kernel $L \in E^* \hat{\otimes}_\pi F$ such that $\varphi(L) = \mathcal{L}$. Denote the space of these p-summable operators by $L^{(p)}(E, F)$. Grothendieck showed in $[131]$ that this space is a complete metrizable topological space if one introduces on it the metric which is induced by the metric S_p originally defined on the space of p-summable Fredholm kernels:

$$S_p(L) : = \inf \sum_i |\lambda_i|^p \; , \tag{A11}$$

where the infimum is taken over all representations of L in the form (A10).

A.5. The order of a nuclear operator and its Fredholm determinant

Let $L \in E \hat{\otimes}_\pi F$ be a Fredholm kernel. Consider the lower bound q of all real numbers p, $0 < p \leq 1$, such that L is p-summable. The number q is called the order of the Fredholm kernel. The set of all Fredholm kernels of order $\leq q$ is denoted by $E \overset{[q]}{\otimes} F$.

In analogy an operator $\mathcal{L}: E \rightarrow F$ is called nuclear of order q if there exists a Fredholm kernel L of order q with $\mathcal{L} = \varphi(L)$. The set of all nuclear operators of order $\leqslant$ q will be denoted by $L^{[q]}(E, F)$. For nuclear operators of order zero in particular one has the following characterisation:

A nuclear operator $\mathcal{L}$: $E \rightarrow F$ is of order zero if, and only if it has a representation as given in (A10) where the sequence $\{\lambda_i\}$ belongs to the space l_p for all $p > 0$.

Other properties of nuclear operators are the following ones:

Let $\mathcal{L}$: $E \rightarrow F$ be nuclear of order q and let T_1: $F \rightarrow G$ respectively T_2: $G \rightarrow E$ be linear bounded mappings. Then the mapping $T_1 \cdot \mathcal{L} \cdot T_2$: $G \rightarrow G$ is also nuclear of order q.

Furthermore the tensor product of two nuclear operators of order q in the projective topological tensor product of the corresponding spaces is again nuclear of order q.

Finally, Grothendieck proved the following interesting Theorem [95]:

__Theorem A.2__ Let $\mathcal{L} \in L^{[p]}(E, E)$ with $0 \leqslant p \leqslant 2/3$. Then the Fredholm determinant det$(1 - z\mathcal{L})$ is an entire function of order $\leqslant$ r, where $1/r = 1/p - 1/2$, and of genus 0. The operator $\mathcal{L}$ is of trace class and one has

$$\det(1 - z\mathcal{L}) = \prod_{\{i\}} (1 - z\,\lambda_i)$$

respectively

$$\text{trace } \mathcal{L} = \sum_{\{i\}} \lambda_i \, ,$$

where $\{\lambda_i\}$ are the non-vanishing eigenvalues of $\mathcal{L}$ counted according to their algebraic multiplicities. For these eigenvalues one has furthermore

$$\sum_{\{i\}} |\lambda_i|^p < \infty \, .$$

APPENDIX B.

COMPOSITION OPERATORS ON BANACH SPACES OF HOLOMORPHIC FUNCTIONS

We are going to prove in this Appendix a version of a theorem
about composition operators on the Banach space of holomorphic functions over open domains in the space l_1 which is stronger than the one
we gave in [117] . By restricting the discussion to the Banach space l_1
we can weaken quite a lot the technical assumptions we had to make there
for a general Banach space. This is related to the existence of a mo-
notonic Schauder basis in the space l_1 [132] .

Theorem B.1 Let D be an open bounded region in l_1 and z_0 a point in
D. Let $\psi_0 : l_1 \longrightarrow l_1$ be a nuclear mapping of order zero with $\| \psi_0 \| < 1$
such that the mapping $\psi: l_1 \longrightarrow l_1$ defined as $\psi(z) = z_0 + \psi_0(z)$
maps D strictly inside itself, that means $\overline{\psi(D)} \subset D$. Let $f \in A_\infty(D)$ be a holomorphic function on D which is continuous on $\bar{D}$. Define the composi-
tion operator $T : A_\infty(D) \longrightarrow A_\infty(D)$ by

$$T f (z) := \psi(z) \quad f \bullet \psi(z) .$$

Then we have:

1) ψ has exactly one fixed point z^* in D.

2) T is nuclear of order zero.

3) trace $T = \det(1 - \psi_0)^{-1} \psi(z^*) .$

Proof: The proof of this theorem is similar to the one given in [117]
in the more general case. We will use here the fact that l_1 has a monotone Schauder basis for which we take the following vectors e_i, $i \in \mathbb{N}$:

$$(e_i)_k := \delta_{ik} \quad \text{for all } k \in \mathbb{N} .\tag{B1}$$

Then any $z \in l_1$ can be written as

$$z = \sum_{i=1}^{\infty} z_i \, e_i \qquad\qquad (B2)$$

and we have $\|z\| = \sum_{i=1}^{\infty} |z_i|$.

Assertion 1) of the above theorem is a special case of the Earle-Hamilton fixed point Theorem [133] for holomorphic mappings in an arbitrary complex Banach space.

We next prove assertion 2). For this we remark that there exists a ball $K_\delta(z^*)$ of radius δ around the fixed point z^* such that $\overline{\psi(K_\rho(z^*))}$ $\subset K_\delta(z^*)$. This follows immediately from the assumption $|\psi_0| < 1$:

because

$$\psi(z) = z^* + \psi_0(z - z^*) \qquad\qquad (B3)$$

ψ contracts a neighbourhood of z^*. We can assume that D is in fact already a ball around the fixed point z^*. The general case can always be reduced to this case as shown in [117].

Take then any $g \in A_\infty(D)$ and let its Taylor series in D be given by [115]

$$g(z) = \sum_{k=0}^{\infty} 1/k! \; D^k g(z^*) \, (z - z^*)^k. \qquad\qquad (B4)$$

Because $\psi(z^*) = z^*$ and $\overline{\psi(D)} \subset D$ one gets using relation (B3)

$$g \circ \psi(z) = \sum_{k=0}^{\infty} 1/k! \; D^k g(z^*) \, (\psi_0(z - z^*))^k . \qquad\qquad (B5)$$

Since ψ_0 is nuclear of order zero it has the representation

$$\psi_0 = \sum_i \lambda_i \; e_i^* \otimes e_i \qquad\qquad (B6)$$

with $e_i^* \in l_1^*$, $\|e_i^*\| \leqslant 1$, and $\{e_i\}$ the Schauder basis of l_1 as defined in (B1). The sequence $\{\lambda_i\}$ belongs to the space l_p for any $p > 0$, that means

$$\sum_{i=1}^{\infty} |\lambda_i|^p < \infty \ .$$

Inserting representation (B6) into relation (B5) and taking into account the linearity and symmetry properties of the operators $D^k g(z^*)$ we get

$$g \circ \psi(z) = \sum_{k=0}^{\infty} \sum_{\underline{\alpha}, |\underline{\alpha}|=k} 1/\underline{\alpha}! \ \underline{\lambda}^{\underline{\alpha}} \ \underline{e}^*(z-z^*)^{\underline{\alpha}} \ D^k g(z^*) \ (\underline{e}^{\underline{\alpha}}), \qquad (B7)$$

where we used the following notations:

$\underline{\alpha}$ is a multiindex $\underline{\alpha} = (\alpha_i)_{i \in \mathbb{N}}$ with $\alpha_i \in \mathbb{N} \cup \{0\}$ for all i and $|\underline{\alpha}| : =$

$\sum_{i=1}^{\infty} \alpha_i < \infty$.

$$\underline{\lambda}^{\underline{\alpha}} : = \prod_{i \in \mathbb{N}} \lambda_i^{\alpha_i} \ , \ \underline{e}^*(z-z^*)^{\underline{\alpha}} : = \prod_{i \in \mathbb{N}} e_i^*(z-z^*)^{\alpha_i} \quad \text{and} \qquad (B8)$$

$$\underline{e}^{\underline{\alpha}} : = (\overbrace{e_1, \ldots, e_1}^{\alpha_1}, \ldots, \overbrace{e_i, \ldots, e_i}^{\alpha_i}, \ldots) \in l_1^{|\underline{\alpha}|} \ .$$

This allows us to write the operator $\tilde{T} f (z) : = f \circ \psi(z)$ as

$$\tilde{T} = \sum_{k=0}^{\infty} \sum_{\underline{\alpha}, |\underline{\alpha}|=k} \lambda_{k,\underline{\alpha}} \ \tilde{e}^*_{k,\underline{\alpha}} \otimes \tilde{e}_{k,\underline{\alpha}} \qquad (B9)$$

with

$$\lambda_{k,\underline{\alpha}} : = \prod_{i \in \mathbb{N}} \lambda_i^{\alpha_i \varepsilon} \ ,$$

$$\tilde{e}^*_{k,\underline{\alpha}}(g) : = 1/\underline{\alpha}! \ D^k g(z^*) \ (\underline{e}^{\underline{\alpha}}) \in A_{\infty}(D)^* \ , \qquad (B10)$$

$$\tilde{e}_{k,\underline{\alpha}}(z) : = \underline{e}^*(z-z^*)^{\underline{\alpha}} \ \underline{\lambda}^{\underline{\alpha}(1-\varepsilon)} \in A_{\infty}(D) \ .$$

The number ε is chosen in such a way that

$$\| \sum_i \lambda_i^{1-\varepsilon} e_i^*(z-z^*) \ e_i \| \le \delta'' < \delta \qquad \text{for all } z \in \bar{D} \ . \qquad (B11)$$

This is possible because

$$\left\| \sum_{\{i\}} \lambda_i \, e_i^*(z-z^*) \, e_i \right\| \leq \delta' < \delta \qquad \text{for all } z \in \bar{D} .$$

Relation (B11) is obviously equivalent to

$$\sum_i |\lambda_i|^{1-\varepsilon} \, |e_i^*(z-z^*)| \leq \delta'' . \tag{B12}$$

Next we prove a lemma.

Lemma B.1 Let $\underline{\alpha}$ be a multiindex with $|\underline{\alpha}| = k$ and let α_{ϱ_i}, $1 \leq i \leq s$ be the non-vanishing entries of $\underline{\alpha}$. Then there exist numbers $t_i > 0$, $1 \leq i \leq s$, with $\sum_{i=1}^s t_i < \delta$ such that

1) $\|\tilde{e}_{k,\underline{\alpha}}\| \leq \prod_{i=1}^s t_i^{\alpha_{\varrho_i}}$,

2) $\|\tilde{e}_{k,\underline{\alpha}}^*\| \leq \prod_{i=1}^s t_i^{-\alpha_{\varrho_i}}$.

Proof: Denote by $K_{\delta''} \subset \mathbb{R}^s$ the following compact set

$$K_{\delta''} = \left\{ x = (x_1,\ldots,x_s) \subset \mathbb{R}^s : \ x_i \geq 0 \ \forall \, i \ \text{and} \ \sum_{i=1}^s x_i \leq \delta'' \right\}.$$

Let $\eta : K_{\delta''} \to \mathbb{R}$ be the continuous function

$$\eta(x) := \prod_{i=1}^s x_i^{\alpha_{\varrho_i}} . \tag{B13}$$

Since η is continuous there exists a $x_o \in K_{\delta''}$ with $\eta(x_o) \geq \eta(x)$ for all $x \in K_{\delta''}$. Let $x_o = (t_1,\ldots,t_s)$. Because $\sum_{i=1}^s |\lambda_{\varrho_i}|^{1-\varepsilon} \, |e_{\varrho_i}^*(z-z^*)| \leq \delta''$ for all $z \in \bar{D}$ we get

$$\prod_{i=1}^s |\lambda_{\varrho_i}|^{(1-\varepsilon)\alpha_{\varrho_i}} \, |e_{\varrho_i}^*(z-z^*)|^{\alpha_{\varrho_i}} \leq \prod_{i=1}^s t_i^{\alpha_{\varrho_i}} . \tag{B14}$$

But this proves just our assertion 1) of Lemma B.1.

Let $\tilde{e}_{k,\underline{\alpha}}^*(g) = \prod_{i=1}^s 1/\alpha_{\varrho_i}! \; D^k g(z^*) \, (e_{\varrho_1},\ldots,e_{\varrho_1},\ldots,e_{\varrho_s},\ldots,e_{\varrho_s}) .$

The right hand side of this expression is nothing else but

$$1/(2\pi i)^s \quad \frac{\partial^k}{\partial z_1^{\alpha\rho_1}\ldots\partial z_s^{\alpha\rho_s}} \quad g(z^* + \sum_{i=1}^{s} z_i \, e_{\rho_i}) \Big|_{z_i=0} \, .$$

We can therefore apply Cauchy's inequalities to the function $\tilde{g}$ in the polycylinder $P = \{z \in \mathbb{C}^s : |z_i| < t_i\}$ which because of the inequality $\sum_{i=1}^{s} t_i \leq \delta''$ belongs completely to the domain of holomorphy of this function. The function $\tilde{g}$ is thereby defined as

$$\tilde{g}(z_1,\ldots,z_s) := g(z^* + \sum_{i=1}^{s} z_i \, e_{\rho_i}) \, .$$

Doing this we get

$$\Big| 1/(2\pi i)^s \quad \frac{\partial^k}{\partial z_1^{\alpha\rho_1}\ldots\partial z_s^{\alpha\rho_s}} \tilde{g}(\vec{z}) \Big|_{\vec{z}=0} \quad \leq \sup_{z \in \bar{D}} |g(z)| / t_1^{\alpha\rho_1}..t_s^{\alpha\rho_s} \, .$$

But this completes the proof of assertion 2) of Lemma B.1.

Coming back to the proof of our Theorem B.1 we introduce the quantities

$$e_{k,\underline{\alpha}} := (\prod_{i=1}^{s} t_i^{\alpha\rho_i})^{-1} \tilde{e}_{k,\underline{\alpha}}$$

$$\tilde{e}_{k,\underline{\alpha}}^* := (\prod_{i=1}^{s} t_i^{\alpha\rho_i}) \, e_{k,\underline{\alpha}}^* \, .$$

(B15)

In terms of them the operator $\tilde{T}$ can then be written as

$$\tilde{T} = \sum_{k=0}^{\infty} \sum_{\underline{\alpha},|\underline{\alpha}|=k} \lambda_{k,\underline{\alpha}} \, e_{k,\underline{\alpha}}^* \otimes e_{k,\underline{\alpha}}$$

(B16)

with

$$\sum_{k=0}^{\infty} \sum_{\underline{\alpha},|\underline{\alpha}|=k} |\lambda_{k,\underline{\alpha}}|^q < \infty \quad \text{for all } q > 0 \, ,$$

$$e_{k,\underline{\alpha}}^* \in A_\infty(D)^* \quad \text{with } \|e_{k,\underline{\alpha}}^*\| \leq 1 \quad \text{and}$$

$$e_{k,\underline{\alpha}} \in A_\infty(D) \quad \text{with} \quad \|e_{k,\underline{\alpha}}\| \leqslant 1 .$$

But this shows that the operator $\widetilde{T}$ is a nuclear operator of order ze-ro. Since the operator T on the other hand is the composition of this operator with a bounded linear multiplication operator also T is nuc-lear of order zero in the space $A_\infty(D)$.

Assertion 3) of Theorem B.1 finally is proved exactly in the same way as the analogous trace formula in reference [117] so that we can omit the details here.

APPENDIX C.

POSITIVE OPERATORS IN BANACH SPACES

In this Appendix we collect some important results of the theory of
positive operators in general Banach spaces relevant for our work here.
A detailed discussion of this theory can be found in [110] and [134].

Let B be a real Banach space. A subset $K \subset B$ is called a proper
cone if

(K1) with $x \in K$ also $\rho x \in K$ for all $\rho \geqslant 0$,

$$(C1)$$

(K2) if $x \in K$ and $-x \in K$, then $x = 0$.

Let K be a proper cone. K is called reproducing if every $z \in B$ can
be written as $z = x - y$, with $x, y \in K$, that means, if $B = K - K$.
Every proper cone induces a partial order $\leqslant$ in B : let $x, y \in B$. Then
we say

$$x \leqslant y \quad \Longleftrightarrow \quad y - x \in K .$$

$$(C2)$$

A linear operator $T : B \longrightarrow B$ is called a positive operator if T
leaves the cone K invariant: $T K \subset K$.

Let $u_o \in K$, $u_o \neq 0$.

<u>Definition C.1</u> A positive operator T is called u_o-positive if there
exist for every $x \in K$, $x \neq 0$ a number $p \in \mathbb{N}$ and positive real numbers
$\alpha, \beta > 0$ such that

$$\beta u_o \leqslant T^p x \leqslant \alpha u_o .$$

This class of positive operators has been extensively studied by Krasnoselskii and he showed [109], [110], [134] that these operators allow for a generalization of the Theorems of Perron-Frobenius [76-77]
respectively of Jentzsch [84].

<u>Theorem C.1</u> (Krasnoselskii) Let K be a reproducing cone in a real Banach space B. Assume $T : B \longrightarrow B$ to be an u_o-positive compact linear operator in B. Chose $p \in \mathbb{N}$ and $\alpha, \beta > 0$ such that $\beta u_o \leq T^p u_o \leq \alpha u_o$. Then one has:

1) There exists an eigenvector x_1 unique up to scalar multiplication in the cone K with $T x_1 = \lambda_1 x_1$. The eigenvalue λ_1 is strictly positive and can be estimated by

$$\beta^{1/p} \leq \lambda_1 \leq \alpha^{1/p}.$$

2) The eigenvalue λ_1 is simple and all other eigenvalues of T (considered as a complex linear operator in the complexified Banach space $B_{\mathbb{C}}$) are in absolute value strictly smaller than λ_1 .

It is obvious that this theorem reproduces for finite dimensional Banach spaces just the results of Perron and Frobenius and for integral operators on $\ell(M)$ where M is some compact manifold, the result of Jentzsch.

How can one see if a positive operator is in fact u_o-positive? An answer to this question is given by [110]

<u>Lemma C.1</u> Let T be a positive operator and let $u_o \in K$, $u_o \neq 0$. If there exist natural numbers q and p and real numbers $\alpha, \beta > 0$ such that

$$T^p x \geq \beta u_o \quad \text{respectively} \quad T^q x \leq \alpha u_o,$$

then T is already u_o-positive.

A simple application of this theory for certain composition operators in complex Banach spaces of holomorphic functions can be found in [135] . Here we recall only the most important result of this work.
To formulate it we need some definitions. Let $D \subset \mathbb{C}^n$ be an open bounded domain in $\mathbb{C}^n$. Let $A_\infty(D)$ be the Banach space of all holomorphic functions on D with the sup-norm. We denote by $H_{in}(D)$ the set of all holomorphic mappings $\psi : D_1 \longrightarrow D$ where D_1 is some small open neighbourhood of $\bar{D}$. It is then known that ψ has exactly one fixed point z^* in

D [133]. Define a set $D_{\mathbb{R}}(z^*)$ as

$$D_{\mathbb{R}}(z^*) \quad := D \cap \left\{ z^* + \mathbb{R}^n \right\} . \qquad (C3)$$

Consider then mappings $\psi \in H_{in}(D)$ with

$$\overline{\psi(D_{\mathbb{R}}(z^*))} \subset D_{\mathbb{R}}(z^*) . \qquad (C4)$$

This is just a certain reality condition on the mapping ψ. The set of all mappings $\psi \in H_{in}(D)$ which fulfil (C4) we denote by $H_{in}^{\mathbb{R}}(D)$. Let $\psi_k \in H_{in}^{\mathbb{R}}(D)$, $1 \le k \le m$, such that there exists a k_0, $1 \le k_0 \le m$, with $z_k^* \in D_{\mathbb{R}}(z_{k_0}^*)$ for all k.

Let $\underline{\alpha} = (\alpha_i)_{i \in \mathbb{N}}$ be a multiindex with $\alpha_i \in \mathbb{N} \cup \{0\}$ and $|\underline{\alpha}| = \sum_{i=1}^{\infty} \alpha_i < \infty$. We denote by $\psi^{\underline{\alpha}_1, \cdots, \underline{\alpha}_m}$ the following mapping of D into D:

$$\psi^{\underline{\alpha}_1, \cdots, \underline{\alpha}_m} : = \psi_1^{\alpha_{11}} \circ \cdots \circ \psi_m^{\alpha_{m1}} \circ \psi_1^{\alpha_{12}} \circ \cdots \circ \psi_m^{\alpha_{m2}} \circ \cdots \circ \psi_1^{\alpha_{1i}} \circ \cdots \circ \psi_m^{\alpha_{mi}} \circ \cdots \qquad (C5)$$

Obviously the mapping $\psi^{\underline{\alpha}_1, \cdots, \underline{\alpha}_m}$ is again in $H_{in}^{\mathbb{R}}(D)$.

Let $f \in A_\infty(D)$ and let Δ be a domain with $\overline{\Delta} \subset D$. Let $G_1, \ldots, G_r$ be the (2n-2)-dimensional analytic null sets of f in $\overline{\Delta}$ [137].

Definition C.2   The mappings $\psi_1, \ldots, \psi_m \in H_{in}^{\mathbb{R}}(D)$ are called separating, if

1) there exists a k_0, $1 \le k_0 \le m$ with $z_k^* \subset D_{\mathbb{R}}(z_{k_0}^*)$ for all $1 \le k \le m$,

2) for all regions Δ with $\overline{\Delta} \subset D$ there exists a number $N_0 < \infty$ such that for all $N > N_0$ and all $z \in \overline{\Delta} \cap \overline{D}_{\mathbb{R}}$ for which there exists a multiindex $\underline{\alpha}$, $\underline{\alpha} = (\underline{\alpha}_1, \ldots, \underline{\alpha}_m)$ with $|\underline{\alpha}| = N$ and $\psi^{\underline{\alpha}_1 \cdots \underline{\alpha}_m}(z) \in \bigcup_i G_i \cap D_{\mathbb{R}}$ there exists another multiindex $\underline{\beta} = (\beta_1, \ldots, \beta_m)$ with $|\underline{\beta}| = N$ and $\psi^{\underline{\beta}_1 \cdots \underline{\beta}_m}(z) \notin G_i \cap D_{\mathbb{R}}$ for any $1 \le i \le r$.

Condition 2) in the above definition just says that the set of points $\left\{ \psi^{\underline{\alpha}_1, \cdots, \underline{\alpha}_m}(z) \right\}$ is so dense in $D_{\mathbb{R}}$ for every z that there does not exist any function $f \in A_\infty(D)$ whose null sets contain all these points without being identically zero.

After this bulk of definitions we can finally formulate the main result of our investigations in [135]:

<u>Theorem C.2</u> Let $D \subset \mathbb{C}^n$ be an open bounded simply connected domain. Assume the mappings $\psi_1, \ldots, \psi_m \in H_{in}^{\mathbb{R}}(D)$ separating. Let $\varphi_k \in A_\infty(D_1)$ for $1 \le k \le m$, where D_1 is some open neighbourhood of $\bar{D}$, with $\varphi_k|_{\bar{D}_{\mathbb{R}}} > 0$. Define a mapping $T : A_\infty(D) \to A_\infty(D)$ as $Tf(z) = \sum_{k=1}^{m} \varphi_k(z) \; f \circ \psi_k(z)$. Then there exists an eigenvalue λ_1 of the operator T with $\lambda_1 > 0$ and $\lambda_1 > |\lambda|$ for all other eigenvalues $\lambda \ne \lambda_1$ of T. The corresponding eigenfunction f_1 is strictly positive on $D_{\mathbb{R}}$ and one has furthermore for λ_1 the estimates

$$\max_{z \in \bar{D}_{\mathbb{R}}} \sum_{k=1}^{m} \varphi_k(z) \ge \lambda_1 \ge \min_{z \in \bar{D}_{\mathbb{R}}} \sum_{k=1}^{m} \varphi_k(z).$$

Bibliography

[1] E.Lieb, D.Mattis: Mathematical Physics in One Dimension.
Academic Press, New York (1966).

[2] C.Thompson: One Dimensional Models-Short Range Forces. In "Phase
Transitions and Critical Phenomena", vol.1, eds. C.Domb,
M.Green, Academic Press, London (1972).
"Mathematical Statistical Mechanics". The Macmillan Com-
pany, New York (1972).

[3] Z.Salsburg, R.Zwanzig, J.Kirkwood: Molecular Distribution Functions in a One Dimensional Fluid. J. Chem. Phys. $\underline{21}$,
1098-1107 (1953).

[4] L.Kadanoff: Scaling Laws for Ising Models Near T_c. Physics $\underline{2}$,
263-272 (1966).

[5] K.Wilson, I.Kogut: The Renormalization Group. Phys. Rep. $\underline{12C}$,
75-200 (1974).

[6] P.Bleher, Y.Sinai: Investigations of the Critical Point in Models
of the Type of Dysons Hierarchical Model. Commun. Math.
Phys. $\underline{33}$, 23-42 (1973).

[7] P.Collet, J.Eckmann: A Renormalization Group Analysis of the
Hierarchical Model in Statistical Mechanics. Lect. Notes
in Physics, no. $\underline{74}$, Springer Verlag, Berlin (1978).

[8] J.Glimm, A.Jaffe: Quantum Field Theory Models, in "Statistical
Mechanics and Quantum Field Theory ", eds. C.de Witt,
R.Stora, Gordon and Breach, New York (1971).

[9] J.Glimm, A.Jaffe: A Tutorial Course in Constructive Field Theory,
in "New Developments in Quantum Field Theory and Statistical Mechanics", eds. M.Levy, P.Mitter, Plenum Press,
New York (1977).

[10] Y.Sinai: Markov Partitions and C-Diffeomorphisms. Funct. Anal.
Appl. $\underline{2}$, no 1, 64-89 (1968).

[11] Y.Sinai: Construction of Markov Partitions. Funct. Anal. Appl. $\underline{2}$,
        no 2, 70-80 (1968).

[12] Y.Sinai: Gibbs Measures in Ergodic Theory, Russ. Math. Surv. $\underline{166}$,
        21-69 (1972).

[13] R.Bowen: Markov Partitions for Axiom A Diffeomorphisms. Am. J.
        Math. $\underline{92}$, 725-747 (1970).

[14] R.Bowen: Markov Partitions and Minimal Sets for Axiom A Diffeo-
        morphisms. Am. J. Math. $\underline{92}$, 907-918 (1970).

[15] R.Bowen: Periodic Points and Measures for Axiom A Diffeomorphisms.
        Trans. Am. Math. Soc. $\underline{154}$, 377-397 (1971).

[16] R.Bowen: Symbolic Dynamics for Hyperbolic Flows. Am. J. Math. $\underline{95}$,
        429-459 (1973).

[17] R.Bowen: Some Systems with Unique Equilibrium States. Math. Syst.
        Theory $\underline{8}$, no 3, 193-202 (1974).

[18] R.Bowen: Bernoulli Equilibrium States for Axiom A Diffeomorphisms.
        Math. Syst. Theory $\underline{8}$, 289-294 (1975).

[19] R.Bowen: Equilibrium States and the Ergodic Theory of Anosov
        Diffeomorphisms. Lect. Notes in Math., vol. $\underline{470}$, Springer
        Verlag, Berlin (1975).

[20] R.Bowen, D.Ruelle: The Ergodic Theory of Axiom A Flows. Invent.
        Math. $\underline{29}$, 181-202 (1975).

[21] D.Ruelle: A Measure Associated with Axiom A Attractors. Am. J.
        Math. $\underline{98}$, 619-654 (1976).

[22] D.Ruelle: "Thermodynamic Formalism". Addison-Wesley, Reading,Mass.
        (1978).

[23] J.Lebowitz: Ergodic Theory and Statistical Mechanics. In "Trans-
        port Phenomena", eds. G.Kirczenow, J.Marro, Lect. Notes
        in Physics, vol. $\underline{31}$, Springer Verlag, Berlin (1974).

[24] J.Moser: Dynamical Systems, Theory and Applications. Lect. Notes
        in Physics, vol. $\underline{38}$, Springer Verlag, Berlin (1975).

[25] S.Smale: Differentiable Dynamical Systems. Bull. Am. Math. Soc.

$\underline{73}$, 747-817 (1967).

[26] E.Lorenz: Deterministic Nonperiodic Flow. J. Atmos. Sci. $\underline{20}$, 130-141 (1963).

[27] D.Ruelle, F.Takens: On the Nature of Turbulence. Commun. Math. Phys. $\underline{20}$, 167-192 (1971), Commun. Math. Phys. $\underline{23}$, 343-344 (1971).

[28] D.Ruelle: The Lorenz Attractor and the Problem of Turbulence. Report at the Conference on "Quantum Models and Mathematics", Bielefeld (1975).

[29] F.Schlögl: Chemical Reaction Models for Non-Equilibrium Phase Transitions. Zeitschrift für Physik $\underline{253}$, 147-161 (1972).

[30] O.Lanford III: CIME Summer Course in Statistical Mechanics, Bressanone, Italy (1976).

[31] G.Rushbrooke, H.Ursell: On One-Dimensional Regular Assemblies. Proc. Camb. Phil. Soc. $\underline{44}$, 263-271 (1948).

[32] M.Baur, L.Nosanow: Phase Transitions in One-Dimensional Order-Disorder Systems: Application to Helix-Random-Coil Transition in Polymers. J. Chem. Phys. $\underline{37}$, 153-160 (1962).

[33] L.van Hove: Sur l'Intégrale de Configuration pour les Systemes des Particules à Une Dimension. Physica $\underline{16}$, 137-143 (1950).

[34] D.Ruelle: "Statistical Mechanics, Rigorous Results". Benjamin, New York (1969).

[35] F.Dyson: Existence and Nature of Phase Transitions in One-Dimensional Ising Ferromagnets. In "Mathematical Aspects of Statistical Mechanics". Am. Math. Soc., Providence, R.I.(1972).

[36] E.Ising: Beitrag zur Theorie des Ferromagnetismus. Zeitschrift für Physik $\underline{31}$, 253-258 (1925).

[37] V.Kolomytsev, A.Rokhlenko: Sufficient Conditions for Ordering of an Ising Ferromagnet. Teor. Mat. Fiz. $\underline{35}$, No 3, 322-331 (1978).

[38] P.Anderson, G.Yuval: Exact Results in the Kondo Problem: Equiva-

lence to a Classical One-Dimensional Coulomb Gas. Phys. Rev. Lett. $\underline{23}$, 89-92 (1969).

[39] P.Anderson, G.Yuval, D.Hamann: Scaling Theory for the Kondo and One-Dimensional Ising Models. Solid State Commun. $\underline{8}$, 1033 - 1037 (1970).

[40] F.Dyson: An Ising Ferromagnet with Discontinuous Long Range Order. Commun. Math. Phys. $\underline{21}$, 269-283 (1971).

[41] P.Ehrenfest: Phasenumwandlungen im üblichen und erweiterten Sinn, klassifiziert nach den entsprechenden Singularitäten des Thermodynamischen Potentials. In "P.Ehrenfest, Collected Papers", ed. M.Klein, North Holland, Amsterdam, Netherlands (1959).

[42] G.Gallavotti, F.Lin: One-Dimensional Lattice Gases with Rapidly Decreasing Interactions. Arch. Rat. Mech. and Analys. $\underline{37}$, 181-191 (1970).

[43] F.Dyson: Existence of a Phase Transition in a One-Dimensional Ising Ferromagnet. Commun. Math. Phys. $\underline{12}$, 91-107 (1969).

[44] F.Dyson: Non-Existence of Spontaneous Magnetisation in a One-Dimensional Ising Ferromagnet. Commun. Math. Phys. $\underline{12}$, 212-215 (1969).

[45] M.Fisher: The Theory of Condensation and the Critical Point. Physics $\underline{3}$, 255-283 (1967).

[46] H.Araki: Gibbs States of a One-Dimensional Quantum Lattice. Comm. Math. Phys. $\underline{14}$, 120-157 (1969).

[47] D.Ruelle: Equilibrium Statistical Mechanics of One-Dimensional Classical Lattice Systems. In "International Symposium on Mathematical Problems in Theoretical Physics", ed. H. Araki, Lecture Notes in Physics, vol $\underline{39}$, Springer Verlag, Berlin (1975).

[48] R.Dobrushin: Analyticity of Correlation Functions in One-Dimensional Classical Systems with Slowly Decreasing Potentials.

Commun. Math. Phys. <u>32</u>, 269-289 (1973).

[49] H.Kramers, G.Wannier: Statistics of the Two-Dimensional Ferromagnet. Part I. Phys. Rev. <u>60</u>, 252-262 (1941).

[50] E.Montroll: Statistical Mechanics of Nearest Neighbour Systems. J. Chem. Phys. <u>9</u>, 7o(-721 (1941).

[51] D.Ruelle: Statistical Mechanics of a One-Dimensional Lattice Gas. Commun. Math. Phys. <u>9</u>, 267-278 (1968).

[52] G.Gallavotti, S.Miracle-Sole: Absence of Phase Transitions in Hard-Core One-Dimensional Systems with Long Range Interactions. J. Math. Phys. <u>11</u>, 147-154 (1969).

[53] T.Kato: "Perturbation Theory for Linear Operators". 2nd ed., ch. 3, 4. Springer Verlag, Berlin (1976).

[54] See [22], Theorem 5.26, p.91.

[55] M.Kac: Mathematical Mechanism of Phase Transitions. In "Brandeis University Summer Institute in Theoretical Physics", vol. 1, Gordon-Breach, New York (1966).

[56] E.Stanley: D-Vector Models or "Universality Hamiltonian": Properties of Isotropically-Interacting D-Dimensional Classical Spins. In "Phase Transitions and Critical Phenomena", vol. 3, eds. C.Domb, M.Green, Academic Press, London (1974).

[57] G.Jameson: "Topology and Normed Spaces". Theorem 28.2. Chapman and Hill, London (1974).

[58] See [57], ch. 15.

[59] See [34], ch. 7.

[60] P.Halmos: "Measure Theory", p. 247. Van Nostrand, New York (1950).

[61] See [22], Proposition 1.4.

[62] R.Dobrushin: The Description of a Random Field by Means of Conditional Probabilities and Conditions of its Regularity. Theory Prob. Applic. <u>13</u>, 197-224 (1968).
Gibbsian Random Fields for Lattice Systems with Pairwise Interactions. Func. Anal. Appl. <u>2</u>, 292-301 (1968).

[63] R.Dobrushin: The Problem of Uniqueness of a Gibbsian Random Field and the Problem of Phase Transitions. Funct. Anal. Appl. 2, 302-312 (1968).

[64] O.Lanford, D.Ruelle: Observables at Infinity and States with Short Range Correlations in Statistical Mechanics. Commun. Math. Phys. 13, 194-215 (1969).

[65] See [22] , ch. 1.6.

[66] See [22] , Theorem 1.8.

[67] J.Lebowitz, O.Penrose: Modern Ergodic Theory. Physics Today 26, no 2, 23-29 (1973).

[68] C.Gruber: On the Definition of Phase Transition. J. Stat. Phys. 14, 81-86 (1976).

[69] See [22] , ch. 3.

[70] G.Gallavotti, S.Miracle Sole: Statistical Mechanics of Lattice Systems. Commun. Math. Phys. 5, 317-323 (1967).

[71] L.Onsager: Crystal Statistics: I. A Two-Dimensional Model with an Order-Disorder Transition. Phys. Rev. 65, 117-149 (1944).

[72] T.Lee, C.Yang: Statistical Theory of Equations of State and Phase Transitions, I, II. Phys. Rev. 87, 404-419 (1952).

[73] F.Dunlop: Zeroes of Partition Functions via Correlation Inequalities. IHES-Preprint, Bures sur Yvette (1977).

[74] B.Souillard: Links Between Decay Properties of Correlations and Analyticity of the Pressure and Correlation Functions. In "International Symposium on Mathematical Problems in Theoretical Physics", ed. H.Araki, Lect. Notes in Physics, vol. 39, Springer Verlag, Berlin (1975).

[75] J.Fröhlich, R.Israel, E.Lieb, B.Simon: Phase Transitions and Reflection Positivity: I. General Theory and Long Range Lattice Models. Commun. Math. Phys. 62, 1-34 (1978).

[76] O.Perron: Zur Theorie der Matrizen. Math. Ann. 64, 248-263 (1907).

[77] G.Frobenius: Über Matrizen aus nicht negativen Elementen. S.-B.

Preuss. Akd. Wiss. Berlin, 456-477 (1912).

[78] F.Ledrappier: Mesures d'Équilibre sur un Réseau. Commun. Math.
Phys. 33, 119-128 (1973).

[79] G.Gallavotti: Ising Model and Bernoulli Schemes in One Dimension.
Commun. Math. Phys. 32, 183-190 (1973).

[80] G.Joyce: Classical Heisenberg Model. Phys. Rev. 155, 478-491
(1966).

[81] J.Rae: The Free Energy of the Classical Heisenberg Model with Anisotropic Interactions. J. Phys. A7, 1349-1359 (1974).

[82] G.Joyce: Exact Results for the One-Dimensional, Anisotropic Classical Heisenberg Model. Phys. Rev. Lett. 19, 581-583(1967).

[83] I.Gohberg, M.Krein: "Introduction to the Theory of Linear Nonself-
adjoint Operators". p. 122. Am. Math. Soc., Providence,
R.I. (1969).

[84] R.Jentzsch: Über Integraloperatoren mit positivem Kern. Crelles
J. Math. 141, 235-244 (1912).

[85] C.Thompson: "Mathematical Statistical Mechanics", p. 127. The
Macmillan Company, New York (1972).

[86] L.Tonks: The Complete Equation of State of One, Two and Three
Dimensional Gases of Hard Elastic Spheres. Phys. Rev. 50,
955-963 (1936).

[87] H.Takahashi: Eine einfache Methode zur Behandlung der Statisti-
schen Mechanik eindimensionaler Substanzen. Proc. Phys.
Math. Soc. Japan, Tokyo 24, 60-63 (1942).

[88] J.Lebowitz, E.Presutti: Statistical Mechanics of Systems of Un-
bounded Spins. Preprint Yeshiva University, New York
(1976).

[89] M.Fisher: The Free Energy of a Macroscopic System. Arch. Rat.
Mech. Anal. 17, 377-410 (1964).

[90] I.Gelfand, N.Vilenkin: "Generalized Functions", vol.4. Academic
Press, New York (1964).

[91] A.Grothendieck: Produits Tensoriels Topologiques et Éspaces Nuc-
          léaires. Ch. II. § 1, no 2, p. 9. Mem. Am. Math. Soc. 16,
          (1955).

[92] D.Mayer, K.Viswanathan: On the Zeta Function of a One-Dimensional
          Classical System of Hard-Rods. Commun. Math. Phys. 52,
          175-189 (1977).

[93] See [53] , ch. 7.

[94] See [83] , p. 164.

[95] See [91] , ch. II. § 1 , no 4 , Corollaire 4.

[96] See [53] , ch. 7.

[97] M.Kac: On the Partition Function of a One-Dimensional Gas. Phys.
          Fluids 2, 8-12 (1959).

[98] G.Baker: One-Dimensional Order-Disorder Model which Approaches a
          Second Order Phase Transition. Phys. Rev. 122, 1477-1484
          (1961).

[99] M.Kac, G.Uhlenbeck, P.Hemmer: On the Van der Waals Theory of the
          Vapor-Liquid Equilibrium. I. Discussion of a One-Dimensio-
          nal Model. J. Math. Phys. 4, 216-228 (1964).
          II. Discussion of the Distribution Functions. J. Math.
          Phys. 4, 229-247 (1964).
          III. Discussion of the Critical Region. J. Math. Phys. 5,
          60-74 (1964).

[100] D.Newman: Equation of State for a Gas with a Weak, Long-Range Po-
          sitive Potential. J. Math. Phys. 5, 1153-1157 (1964).

[101] P.Hemmer, J.Lebowitz: Systems with Weak Long-Range Potentials. In
          "Phase Transitions and Critical Phenomena", vol. 5b, eds.
          C.Domb, M.Green, Academic Press, London (1976).

[102] See remark by M.Kac in [55] .

[103] K.Viswanathan: Statistical Mechanics of a One-Dimensional Lattice
          Gas with Exponential- Polynomial Interactions. Commun.
          Math. Phys. 47, 131-141 (1976).

[104] D.Mayer: On a Zeta Function Related to the Continued Fraction
    Transformation. Bull. Soc. Math. France 104, 195-203(1976).

[105] D.Mayer: The Transfer Matrix of a One-Sided Subshift of Finite
    Type with Exponential Interactions. Lett. Math. Phys. 1,
    335-343 (1976).

[106] D.Ruelle: Zeta-Functions for Expanding Maps and Anosov Flows.
    Invent. Math. 34, 231.242 (1976).

[107] See [22], Theorem 5.26.

[108] M.Krein, M.Rutman: Linear Operators Leaving Invariant a Cone in a
    Banach Space. Transl. Am. Math. Soc. Ser. 1, 10, 199-325
    (1950).

[109] M.Krasnoselskii, L.Ladyzenskii: Structure of a Spectrum of Posi-
    tive Non-Homogeneous Operators. Trudy Moskovskovo Matem.
    ob. 3, (1954).

[110] M.Krasnoselskii:"Positive Solutions of Operator Equations". Ch.2.
    P. Noordhoff, Groningen (1964).

[111] J.Dieudonné: "Foundations of Modern Analysis", p. 209. Academic
    Press, New York (1969).

[112] J.Lebowitz, O.Penrose: Rigorous Treatment of the Van der Waals-
    Maxwell Theory of the Liquid-Vapor Transition. J. Math.
    Phys. 7, 98-113 (1966).

[113] R.Potts: Some Generalized Order-Disorder Transformations. Proc.
    Camb. Phil. Soc. 48, 106-109 (1952).
    C.Domb: Configurational Studies of the Potts Models. J. Phys. A7,
    1335-1348 (1974).
    L.Mittag, M.Stephen: Dual Transformations in Many Component Ising
    Models. J. Math. Phys. 12, 441-450 (1971).

[114] D.Mayer, K.Viswanathan: Statistical Mechanics of One-Dimensional
    Ising and Potts Models with Exponential Interactions.
    Physica 89A, 97-112 (1977).

[115] L.Nachbin: "Topology on Spaces of Holomorphic Mappings". Springer
    Verlag, Berlin (1969).

[116] D.Pizanelli: Applications Analytiques en Dimension Infinie. Bull. Soc. Math. France 96, 181-191 (1972).

[117] D.Mayer: On Composition Operators on Banach Spaces of Holomorphic Functions. To appear in J. Funct. Anal. 35, (1980).

[118] D.Ruelle: Generalized Zeta-Functions for Axiom A Basic Sets. Bull. Am. Math. Soc. 82, 153-156 (1976).

[119] E.Artin, B.Mazur: On Periodic Points. Ann. Math. (2) 81, 82-99 (1965).

[120] P.Walters: Ergodic Theory-Introductory Lectures, p. 18. Lecture Notes in Math., vol. 458, Springer Verlag, Berlin (1975).

[121] F.Landau:"Darstellung und Begründung einiger neuerer Ergebnisse der Funktionentheorie". § 17. Springer Verlag, Berlin (1929).

[122] R.Bowen, O.Lanford: Zeta Functions of Restrictions of the Shift Transformation. Proc. Symp. Pure Math. 14, 43-49 (1970).

[123] See [91] , ch. II, § 2, no 2, Lemme 6 .

[124] M.Schechter: On the Spectra of Operators on Tensor Products. J. Funct. Anal. 4, 95-100 (1969).

[125] See [91] , ch. II, § 1, no 4, Théorème 4.

[126] G.Gallavotti: Funzioni Zeta ed Insiemi Basilari. Accad. Lincei. Rend. Sc. Fis. Mat. e Nat. 61, 309-317 (1976).

[127] B.Felderhof, M.Fisher: Phase Transitions in One-Dimensional Cluster Interaction Fluids. Ann. Physics 58, 176-300 (1970).

[128] L.Schwartz: Produits Tensoriels Topologiques d'Éspaces Vectoriels Topologiques. Éspaces Vectoriels Topologiques Nucléaires. Applications. Séminaire Schwartz 1953/54. Sécretariat Mathématique , Paris (1954).

[129] R.Schatten: "A Theory of Cross Spaces". Princeton Univ. Press, Princeton (1950).

[130] A.Grothendieck: La Théorie de Fredholm. Bull. Soc. Math. France, ch. II. 84, 319-384 (1956).

[131] See [91] , ch. II, § 1, no 1.

[132] See [57] , ch. 30.

[133] C.Earle, R.Hamilton: A Fixed Point Theorem for Holomorphic Map-
pings. In "Global Analysis", Proc. Symp. Pure Math., vol.
XVI, ed. S.Chern, S.Smale. Am. Math. Soc., Providence,
R.I. (1970).

[134] J.Zabreyko et al.: "Integral Equations-a Reference Text". Ch. V,
§ 5. Noordhoff, Leyden (1975).

[135] D.Mayer: Spectral Properties of Certain Composition Operators
Arising in Statistical Mechanics. Commun. Math. Phys. 68,
1-8 (1979).

[136] E.Helfand: Statistical Mechanics of Systems with Long-Range Interactions. In "The Equilibrium Theory of Classical Fluids". Eds. H.Frisch, J.Lebowitz. Benjamin, New York(1964).

[137] H.Behnke, P.Thullen:"Theorie der Funktionen mehrerer komplexer
Veränderlichen". Ch. V. Second Edition, Springer Verlag,
Berlin (1970).

Index

Communications in
Mathematical Physics

ISSN 0010-3616   Title No. 220

Communications in Mathematical Physics is a journal devoted to physics papers with mathematical content. The various topics cover a broad spectrum from classical to quantum physics; the individual editorial sections illustrate this scope:

Subscription information and sample copy upon request.

Springer-Verlag
Berlin
Heidelberg
New York

Selected Issues from
Lecture Notes in Mathematics

Lecture Notes in Physics